CHRISTOPHER SAXTON
ELIZABETHAN MAP-MAKER

Frontispiece *Frontispiece to Saxton's Atlas of England and Wales, 1579*: State II.

CHRISTOPHER SAXTON

ELIZABETHAN MAP-MAKER

by
IFOR M. EVANS
and
HEATHER LAWRENCE

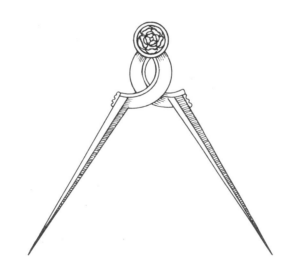

WAKEFIELD HISTORICAL PUBLICATIONS
AND
THE HOLLAND PRESS

MAP-LC
GA
793.5
.S3
E9

First published in 1979
by
Wakefield Historical Publications
&
The Holland Press Limited

© Ifor M. Evans and Heather Lawrence 1979
All rights reserved
Sole distributors in the USA
W. Graham Arader III
1000 Boxwood Court, King of Prussia,
Pennsylvania 19406, USA
(Telephone: 215/825 6570)

ISBN 0 901869 06 6
ISBN 0 900470 95 X

Wakefield Historical Publications
Seckar House, Woolley, Wakefield, West Yorkshire WF4 2LE
Telephone: 0924 257858

The Holland Press Limited
37 Connaught Street, London W2 2AZ
Telephone: 01-262 6184

AUG 1 8 1980

Printed in Great Britain by
The Scolar Press Ilkley, West Yorkshire

Naumburg

Contents

		Page
Frontispiece	Frontispiece to Saxton's Atlas of England and Wales, 1579: *State II*	ii
	List of Plates	vii
	Acknowledgements	ix
	Preface	xi
	Introduction	xiii
Chapter 1	The Saxtons of Dunningley	1
Chapter 2	The National Survey	9
Chapter 3	Saxton's Atlas of England and Wales, 1579, and Wall-map of 1583	20
Chapter 4	Sources of Information and Contemporary Surveying Practices	40
Chapter 5	The Later Editions of Saxton's Atlas	45
Chapter 6	Rewards and Recognition	66
Chapter 7	Christopher Saxton: The Estate Surveyor	74
Chapter 8	Christopher Saxton's Manuscript Maps and Surveys	79
Chapter 9	Robert Saxton's Manuscript Maps and Surveys	122
	Conclusion	139
	Appendices	141
	References	169
	Bibliography	175
	Index	179

Table 1	Saxton's County Maps in the Burghley Atlas	15
Table 2	A Suggested Chronological Sequence for Saxton's County Surveys	18
Table 3	Saxton's Atlas: Titles of Maps	36
Table 4	The Scales of Saxton's County Maps	38
Table 5	The engravers of Saxton's Maps in the Atlas of England and Wales, 1579	39
Table 6	Web's Edition, 1645: Titles of Maps	59
Table 7	Maps in the Saxton Atlas by Lea, *c*.1689	61
Table 8	Maps in the Saxton Atlas by Lea, *c*.1693	62
Table 9	Christopher Saxton's Manuscript Maps and Surveys	79
Table 10	Robert Saxton's Manuscript Maps and Surveys	122

Appendix 1	Extracts from the Lay Subsidies for Ossett and West Ardsley	141
Appendix 2	Woodkirk Parish Register Transcripts relating to Saxton	141
Appendix 3	Extracts from the Feet of Fines 1581–1622	141
Appendix 4	Thomas Saxton's Will	142

		Page
Appendix 5	Robert Saxton's Will	142
Appendix 6	Burghley Atlas	143
Appendix 7	Grant of Land in Suffolk	147
Appendix 8	Order of Assistance for Saxton's Survey of Wales	147
Appendix 9	Licence to Publish	147
Appendix 10	Extant copies of Atlas	148
Appendix 11	Titles and Sequence of Maps	154
Appendix 12	Projected Edition of 1665	155
Appendix 13	Lea's Editions	158
Appendix 14	Later Editions Extant	162
Appendix 15	General Order of Assistance	163
Appendix 16	Grant of Office of Bailiff in London	163
Appendix 17	Grant of Arms	164
Appendix 18	Pedigrees of Early Saxtons	164
Appendix 19	Grant of Land in London	165
Appendix 20	Letter concerning Land in London	165
Appendix 21	Grant of Rectory of Scalby	166
Appendix 22	Appointment as Bailiff of Duchy of Lancaster	166
Appendix 23	Warrant concerning above	167
Appendix 24	Land measurements	167

Plates

		Page
1	Frontispiece to Saxton's Atlas, 1579. Detail of Queen Elizabeth enthroned (*State I*).	21
2	Index to Saxton's Atlas, 1579. *Setting A, Variant II.*	22
3	Index to Saxton's Atlas, 1579. *Setting B.*	23
4	Index to Saxton's Atlas, 1579. *Setting D.*	24
5	Saxton's Atlas, 1579. Plate of coat-of-arms and table of cities.	26
6	Saxton's Atlas, 1579. Map of 'Anglia . . .' *State II.*	27
7	Titlepage to William Web's edition of Saxton's Atlas, 1645.	46
8	Titlepage to Philip Lea's edition of Saxton's Atlas, *c.*1689.	51
9	Saxton's Atlas, 1579. Map of Northamptonshire, etc. *State II.*	54
10	Philip Lea's edition of Saxton's Atlas, *c.*1689. Map of Northamptonshire, etc.	55
11	Philip Lea's edition of Saxton's Atlas, *c.*1689. Map of Northumberland.	56
12	Saxton's map of Northumberland in the process of subsequent modification.	57
13	Manuscript map of Elland Park by Christopher Saxton, 1597.	92
14	Written survey of Thornhill demesne by Christopher Saxton, 1602.	98
15	Manuscript map of Old Byland by Christopher Saxton, 1598.	103
16	Manuscript map of Snapethorpe by Christopher and Robert Saxton, 1601.	113
17	Manuscript map of Baildon by Robert Saxton, 1610.	125
18	Written survey of Esholt by Robert Saxton, 1612.	128
19	Map showing places mentioned in the text and places surveyed within thirty miles of Dunningley in the West Riding of Yorkshire.	138

Acknowledgements

Grateful acknowledgement is made to those many persons who have so kindly contributed to this book. This includes those who have provided details regarding copies of Saxton's Atlas, its later editions and manuscript maps and surveys for which they have responsibility (see Appendix 10 and Tables 9 and 10). Certain locations cannot be named, nor their holdings listed, on the grounds of security; but the contributions of such libraries or private persons have been, despite their anonymity, of considerable usefulness. Thanks are extended to Sir Hugo Boothby for permission to include his atlas in the list of privately held copies.

Sincere thanks are given to the many individuals who have contributed factual information, helped in translating documents, in genealogical research, in supplying information concerning the location of hitherto unknown manuscript maps and surveys, in giving permission to inspect them, in giving advice on specific detail and, by no means least, in the publication of the book. We are particularly grateful to the following: John Andrews, Jane Beavers, J. T. Cliffe, the late Major G. E. Dent, His Grace the Duke of Devonshire, Peter Eden, Kenneth Emsley, Margaret Faull, John Goodchild, J. B. Harley, Richard Knowles, Simon Lawrence, E. Leeson, David Marcombe, David Michelmore, A. D. Mills, His Grace the Duke of Northumberland, Michael Palmer, T. K. Smith, Sir Robert Somerville, David Thomson, Sylvia Thomas and Kate Taylor. Many archivists have also been most helpful; space does not permit inclusion of names but particular thanks go to the archivists and staff of the following record offices: Greater London, Norfolk, Nottinghamshire, Suffolk, Cambridge University, Leeds City, the Borthwick Institute of Historical Research at York, the Brotherton Library at Leeds University and Oxford University. We are also grateful to the staff of the Public Record Office, London, and the Yorkshire Archaeological Society, Leeds, for their help so willingly given. We thank Margery Raisbeck most warmly for her design of the cover.

For permission to reproduce plates for this book and assistance we are indebted to Dr Helen Wallis, Map Librarian at the British Library, London (Plates 1, 2, 3), to Miss Betty Fathers, Map Librarian at the Bodleian Library, Oxford (Plates 4, 5, 6, 9, 10, 12), Lord Savile and the Chief Archivist of Nottinghamshire Record Office (Plates 13, 14 and the drawing of Thornhill), the Yorkshire Archaeological Society, Leeds (Plates 17, 18), the Brotherton Library, Leeds University (Plates 8, 11), Raymond O'Shea, Baynton-Williams Gallery (the frontispiece), the Syndics of Cambridge University Library (Plate 7) and Mr Michael Foljambe (the drawings on page 76). Plate 15 appears by permission of the Controller of Her Majesty's Stationery Office.

Transcripts and translations of records in the Public Record Office appear by permission of the Controller of Her Majesty's Stationery Office and the Chancellor and Council of the Duchy of Lancaster.

Finally we thank Richard Leech for his timely intervention in assuring the publication of the book. Also those who had sufficient faith in the project in its early stages to offer financial support, namely Helen Kilburn, S. T. E. Lawrence, Eric Raper, Mr and Mrs R. H. Baynton-Williams, Raymond Eddy, D. Haigh, Peter Harris, A. C. Moreland, Dennis Morrison, Duncan Mutch, Christopher Rainbow, Michael L. Rakusen, The Regent Gallery Cheltenham, Clifford Stephenson, J. B. Sutton and The Marc Fitch Fund.

Preface

Christopher Saxton's atlas of the counties of England and Wales, published in 1579, ranks as the first national atlas of any country. It did not bear the name atlas. This term to describe a collection of maps was invented by the Flemish geographer Gerard Mercator, and first appeared on the titlepage of his world atlas of 1595. Saxton's atlas had no title or titlepage, but opened with a fine engraved frontispiece depicting Queen Elizabeth I. The Queen thus seemed to preside over the undertaking and invested it with the spirit of Elizabethan England.

It is a measure of Saxton's achievement that only twenty years earlier the Norwich physician William Cunningham had published what he claimed to be the first geographical treatise written in English. In *The Cosmographical Glasse* (1559) he showed how to make a map of England and included a simple diagram marking nineteen English towns on a grid. His map of Norwich which was included in the book is believed to be the first English engraved town plan. Saxton's survey of the counties of England and Wales and publication of the maps in atlas form thus marked a notable advance, and can be considered one of the outstanding cartographic enterprises of the sixteenth century.

That statesmen now felt an urgent need for regional maps of England is indicated by an item in Robert Beale's list of the duties of a Secretary of State, written in 1592. 'A Secretarie must likewise have ... a booke of the Mappes of England with a particular note of the division of the shires into Hundreds, Lathes, Wappen-taes, and what Nobleman, Gentleman and others be residing in ... them.' Lord Burghley as Lord Treasurer to Queen Elizabeth provided himself with thirty-five of Saxton's maps comprising early proofs pulled between 1574 and 1578, and these, heavily annotated, are preserved in the Burghley-Saxton atlas, which came to the British Museum in the old Royal Library. An anecdote told of Lord Burghley illustrates his concern to promote his countrymen's interest in their native land. It is related by Henry Peacham (1622) that 'if anyone came to the Lords of the Council for a licence to travel, he would first examine him of England; if he found him ignorant, would bid him stay at home and know his own country first'. We can consider Saxton's atlas required reading for those who failed to pass the test.

It is appropriate that the 400th anniversary of the appearance of the atlas should be celebrated by the publication of Saxton's biography. The sixteenth century antiquary William Camden called Saxton 'optimus Chorographus', referring to the art of chorography, as the study of regional and local geography was then described. Yet many details of Saxton's life remained unknown. Ifor M. Evans and Heather Lawrence have now filled in many gaps in this welcome biographical study. They have brought to light new facts about the career of Saxton, and have discovered some hitherto unknown maps and surveys. As successor to Edward Lynam, who was responsible for the publication of the British Museum's facsimile edition of the atlas (1936), and to R. A. Skelton, who prepared the facsimile of Saxton's great wall-map (1974), I am honoured to introduce this notable new contribution to the works on Christopher Saxton.

Helen Wallis,
Keeper of Maps,
The British Museum.
September, 1979

Introduction

Christopher Saxton, 'the Father of English Cartography' as he has been called by modern writers, left a legacy of maps of the counties of England and Wales from which succeeding generations of map-makers drew extensively. At first Saxton was simply called an expert surveyor, but rose to be called 'the Queen's surveyor', the highest status a man in his profession could attain.

Saxton's contribution to the development of cartography is unquestionable in that he was the first man to survey and map the counties of England and Wales. The resultant atlas published in 1579, amazingly accurate in detail, survives as testimony to his expertise when surveying techniques and comprehension of the mathematical sciences were still limited.

The Elizabethans' prodigious interest in cartography is demonstrated by the profusion of treatises on surveying techniques compiled for the emergent professors of the science. For the sedentary layman there were topographical works from pens of travellers who supplied geographical information to the reader without the inherent dangers and toil of first hand experience.

For the statesman maps were the tools of trade; Lord Burghley for instance requested maps to be compiled by a variety of men for military, political and administrative purposes, and his collection of Saxton's maps shows evidence of considerable usage. For the traveller Saxton's maps were the first complete large scale delineation of the whole country. Maps of the British Isles and certain areas of significance were already drawn, but available only to a select few.

Cartography was not a new science however. The first English map of Great Britain was that of the Benedictine Matthew Pairs of St Albans in about 1250, followed by the anonymous 'Gough Map' drawn a century later. By the mid fifteenth century the principles of cartography were understood in England, though work was technically still very inaccurate. It was the Italians who excelled both in the craft of surveying and art of engraving, and the invention in Italy of printing maps from engraved copper plates in 1473 heralded an era of superb craftsmanship in map production. As the trade routes shifted north the Italians were superseded by the north Europeans; centres of scientific learning grew up in Germany and the Low Countries and by the 1570s the Dutch and Flemish map-makers had taken precedence in all branches of the trade.

In 1564 Gerard Mercator, the Dutch cartographer, published a detailed wall-map of the British Isles on eight sheets. His friend and countryman Abraham Ortelius was the first to publish an atlas of the world, the Theatrum Orbis Terrarum in 1570, and a supplement of 1573 contained a map of England by Humphrey Lhuyd. That maps of England were being prepared to accompany Holinshed's *Chronicles*, published in 1577 by the Royal printer Reynold Wolfe, is well documented, but Wolfe died before completing the work and no maps survive to indicate how far he had progressed. Thus the field was open and Christopher Saxton, under the patronage of Thomas Seckford, Master of the Queen's Requests, produced the first atlas of the counties of England and Wales, which was the first national atlas of any country. Amongst their contemporaries it was Thomas Seckford that received the credit, but today it is Saxton's name we salute for his magnificent achievement.

In this volume, compiled to coincide with the four hundredth anniversary of the publication of the atlas, it is intended to present all that is currently known of Saxton's life and work. Since 1927 when Sir George Fordham wrote the first modern biographical notice on Saxton, much further research has been undertaken and a considerable amount of fresh material discovered. Many articles have appeared on Saxton's printed atlas and wall-map, and a number of newly discovered manuscript maps have been documented; the authors acknowledge their debt to all previous writers from whose works they have drawn. In researching this work, however, further information on Saxton's early life and a number of hitherto unrecorded manuscript maps and surveys have been discovered, details of which are given.

CHRISTOPHER SAXTON
ELIZABETHAN MAP-MAKER

Pedigree of the Saxton family of Dewsbury and Dunningley

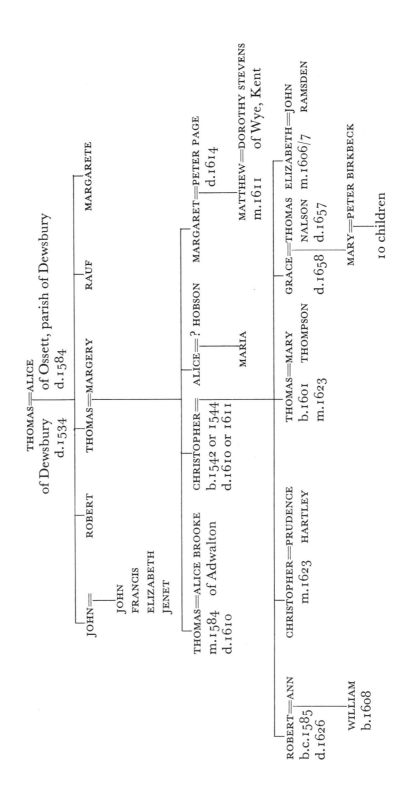

Chapter 1

The Saxtons of Dunningley

Christopher Saxton describes himself as of Dunningley on several occasions. In the sixteenth century there was only a handful of dwellings in Dunningley which is in the parish of West Ardsley, otherwise known as Woodkirk, in the West Riding of Yorkshire. It appears that Saxton was born in the adjoining parish of Dewsbury, probably at Sowood, Ossett, and moved with his family to Dunningley as a child or young man in the period 1545–67. It seems probable that he was educated in Dewsbury where he developed a close association with the vicar, John Rudd, who was a keen cartographer.

The first record of a member of the family living at Dunningley is in 1567 when all men owning goods or property worth over a certain amount were assessed for taxable purposes. In the list of West Ardsley men a Thomas Saxton was assessed and paid 3s 4d; he was also named in subsequent assessments until the year 1599. At the next assessment, in 1606, Christopher Saxton was the only Saxton listed in West Ardsley, and the following time, in 1620–1, Robert Saxton, Christopher's eldest son is named (Appendix 1). The death of Thomas Saxton is recorded in the Woodkirk parish register for 1600 (Appendix 2). In the feet of fines, which record land and property transactions, a fine was levied in 1581 in which Thomas Saxton and his wife Margery were entered as sellers, and a John and Robert Saxton as purchasers of a messuage with lands in 'Woodkirke, Dunynglawe, Tynglawe and Westardeslawe' (Appendix 3). In all probability this Thomas and Margery were Christopher Saxton's parents, but unfortunately no Will can be found for either of them and without further evidence this link must remain unproven.

The name Saxton was comparatively common in the West Riding of Yorkshire particularly in the parishes of Leeds and Dewsbury. A search of Wills surviving at the Borthwick Institute at York has revealed Christopher's immediate ancestry, presuming Thomas to be his father, a hypothesis supported by the recurrence of christian names within the family. (See pedigree page xvi.) A John Saxton of Sawood (Sowood in Ossett) in the parish of Dewsbury, who died in 1587,[1] requested that his brother Thomas of Dunningley be a supervisor of his Will. John had two other brothers, Robert and Rauf and a sister Margarete, and their parents were Thomas Saxton, who died in 1534,[2] when he is described as of Dewsbury, and Alice, who died in 1584,[3] described as of Ossett. Ossett is in the parish of Dewsbury.

Thomas Saxton must have moved to Dunningley sometime between 1545, the date of the lay subsidy in which he is not recorded, and 1567. As Christopher was born in 1542 or 1544 he must have been born in Ossett, or elsewhere in the parish of Dewsbury, and moved to Dunningley between the age of one and twenty-five years. The parishes of Dewsbury and West Ardsley adjoin, so it was not a move of great distance. Sowood, a small hamlet to the south of Ossett, lies four and a half miles south of Dunningley, and almost equidistant from Wakefield and Dewsbury.

Little can be gathered of the status of Christopher's immediate ancestors from the existing Wills. His grandfather Thomas left each of his children £3 6s 8d, the sum of

6s 8d a year to his mother, a foil to John Copley and the residue to his widow. Christopher's brother, Thomas, described himself in his as a yeoman.

Christopher Saxton described himself in 1596 as 'of Dunningley, in the parish of Westardesleye in the Countye of Yorke, Gent., of the age of fiftye twoo yeares or thereabouts'. This would indicate that he was born in 1544, but his inclusion of the qualifying phrase 'or thereabouts' implies that he was not exactly sure. Unfortunately perhaps for the tidy minded historian, he later states, in 1606, that he is aged '64 yeres or thereabouts' which would indicate a date of birth of 1542; once again however, he acknowledges that he is uncertain. If Saxton himself was content to know he was born in 1542, or 1544, 'or thereabouts', then we too must be satisfied.

Christopher Saxton was one of at least four children. His brother Thomas wrote his Will in January 1608 which was proved by his wife Alice in July 1610, some seventeen months later (Appendix 4). In it he mentions his brother Christopher Saxton and his wife, although she is not named. Thomas also mentions Christopher's sons Robert, Christopher and Thomas and daughters 'everyone of them'. His own sisters Alice Hobson and Margaret, wife of Peter Page, and their respective children all received bequests. Thomas and Alice Saxton obviously had no living children of their own as Thomas bequeathed his estate, after small monetary gifts to all named above, to his wife and on her demise to his nephews, Christopher and Thomas Saxton. He must have considered Robert, his eldest nephew, adequately provided for. Thomas described himself as a yeoman of 'Duninglaw' and requested burial in his local parish church at Woodkirk. He had married Alice Brooke at Birstall in 1584, Alice being the daughter of John Brooke of Adwalton.

Nothing can be traced giving Christopher's exact date of death. No Will can be found, nor other contemporary record. There is nothing at the Borthwick Institute at York, the repository for the Wills of those holding land in the Northern Province, where the Wills of other members of his family are found. Nor can one be found in the Wills registered with the Perogative Court of Canterbury, now at the Public Record Office. Where someone such as Saxton had estate worth over £5 held in both jurisdictions the Will would normally have been proved in the higher court, i.e. the Perogative Court of Canterbury.

Christopher Saxton must have died shortly after his brother Thomas, in 1610 or 1611. His last recorded work was in 1608, and his inclusion in his brother Thomas' Will written in January 1608 testify to his being alive at that date. Thomas' Will was proved in June 1610, and in all probability Christopher was still living then. Dr Favour's 'epitaph' to Saxton (see page 6) was written in or by 1611, according to Fordham, which leaves a period of only a few months during which Saxton must have died. He would have been sixty-six, or sixty-eight years old, 'or thereabouts'. It is not known where he was buried, but if he was living at home it must surely have been at Woodkirk with the rest of his family. That Christopher was a Protestant rather than a Catholic is demonstrated by his being appointed churchwarden at Woodkirk in 1605. The obvious source of information, the parish registers of Woodkirk, are missing, as are the Bishop's transcripts, for the years 1602, 1604, 1606, 1607, 1610–22 and 1624–32.

Christopher Saxton's sisters are shadowy characters and throw little light on our study of the man. Alice married a man named Hobson and had a daughter Mary, or Marie, alive in 1626. Margaret Saxton married Peter Page and was living in London in 1589 when the latter was described as a gentleman. Peter Page, son of Thomas Page, came from an old Norfolk family of some standing in his native town of Saxthorpe. The couple were living in the parish of St Giles, Cripplegate, in London, at the time of Peter's death in 1614 and had one son, Matthew, who entered the Church. On his decease Peter Page left considerable property in Suffolk, Cambridgeshire and Bedfordshire to his wife and son, the latter to be installed in the parsonage of Winterton on the Norfolk coast, for which a considerable sum was set aside. A debt was to be settled with his wife Margaret of 'ffower score and odd pounds'; she must have been quite wealthy in her own right. Both Margaret and Matthew Page were alive in 1618, she still living in London. It is quite likely that Saxton lived with the Pages during his long sojourns in London whilst supervising the engraving of his county maps and compiling his wall-map of England and Wales. Peter Page stood surety for Saxton in 1589 when the latter was appointed Bailiff of the Duchy of Lancaster.

Of Christopher's wife nothing is known beyond the fact that she was alive in 1608–10 and mentioned in her brother-in-law Thomas' Will. As Robert, the eldest son, was not born until about 1585, when Christopher would be in his mid forties, it would appear that either she was his second wife or that he married late and did not do so until he had completed his national survey and had a chance to settle down in one place. Whether she was a local Yorkshire girl, or one that he met on his travels is not known. There are few extant parish registers for that date in the area, but the marriage may have taken place in London whilst Saxton was working there. It is probable that Mrs Saxton was considerably younger than her husband as she was still child-bearing in 1601 when Christopher himself was in his late fifties.

The couple had three sons and two daughters living in 1626 when Robert the eldest son wrote his Will. Robert himself was born about 1585 according to his own statement when he surveyed Wensleydale in April 1618, declaring himself to be thirty-three years 'or thereabouts'. He was married before 1608 when his son William was born, and his wife Ann was alive when he wrote his Will in September 1626. The Will, however, was proved by Robert's brother Christopher. In his Will Robert describes himself as a yeoman of the parish of Woodchurch in the County of York, requesting burial in the church there. After monetary bequests to his relatives, his brother Christopher, his brother Thomas and his wife Mary, his sisters Grace Nalson and Elizabeth (Ramsden), his cousin Marie Hobson and her children, and various named godchildren, he bequeathed his sword to his brother-in-law Thomas Nalson. The residue of his estate, lands and 'liveings' he bequeathed to his wife Ann and son William, then a minor (Appendix 5). What became of William has not been discovered, no record of his marriage or death having been traced. It seems unlikely that he had a family; according to Hunter,[4] Grace Saxton, who married Thomas Nalson, was 'heir to her brothers who died without issue'. This is obviously incorrect as evidenced by Robert's son William, but if he died young Grace could well have inherited.[5]

Robert Saxton was Christopher's only son to become a surveyor. There is evidence of his involvement as his father's assistant as early as 1601 when he drew the map of Snapethorpe surveyed by his father. Presumably the two worked together for some time, but Robert was working independently by 1607 when he was commissioned to survey fields in Sandal immediately south of Wakefield. His maps and written surveys, described in Chapter 9, are very similar to his father's manuscript work, but from surviving examples it seems Robert never worked outside the bounds of his native county. His employment came, like that of his father in later years, from both private landlords and the Crown, who engaged him to map or survey controversial boundaries, lands in dispute and estates. Robert and Ann Saxton were involved in several land and property transactions, there being eight recorded in the feet of fines from 1609 to 1622 (Appendix 3).

Little is known of Christopher Saxton's other children, nor is it of much relevance to the cartographic student. His second son Christopher married Prudence Hartley at Woodkirk in 1623 and was alive in 1626 when he was described as of West Ardsley, clothier. Thoms Saxton, Christopher's third son (the order is stated by Thomas senior in his Will) was baptised at Woodkirk on November 22nd, 1601, married Mary Thompson at Woodkirk on November 13th, 1623, and both were alive in 1626. Grace Saxton, reputedly heir to her brothers, married Thomas Nalson of Altofts, she living until September 15th, 1658, and he until November 8th, 1657. Their only daughter Mary married Peter Birkbeck, Rector of Castleford, and they had a family of ten. Elizabeth Saxton married John Ramsden of Greetland at Elland church on February 10th, 1606. John Ramsden was the son of Anthony Ramsden, a yeoman of Greengate Head who had a fulling mill and property in Greetland and Stainland, and his wife Isabel, who was probably the sister of Christopher Cosin, a victim of the notorious Halifax gibbet. In January 1606 Anthony Ramsden entered into a bond of £100 with Christopher Saxton, described as of Dunningley, gent., and John Ramsden, son and heir apparent, concerning conveyance of property on Anthony's death. Elizabeth Saxton, intended wife of John Ramsden, was to be the recipient of some property. The bond was witnessed by Robert Saxton.[6] No children of this marriage have been traced, and none are mentioned in brother Robert's Will of 1626.

Dunningley, the home of Christopher Saxton, is and was a tiny hamlet situated on a hilltop in the parish of Woodkirk almost five miles north-west of Wakefield, four miles south of Leeds and the same from Dewsbury to the south-west. Dunningley Hall, where Saxton lived, was at the northern tip of the hamlet backing on to Dunningley Common; it was demolished in the mid nineteenth century and nothing remains today. Christopher Saxton also owned several fields adjoining and there is later evidence that his son Robert farmed land nearby belonging to Sir John Savile of Howley Hall.[7] The other notable family at Dunningley were the Linleys, classed as yeomen, and both they and the Saxtons were probably clothiers and farmers, the dual economy practised by most of the inhabitants of the district, being situated as it was in the heart of the woollen manufacturing area. Adjoining Dunningley Common was a field called coalpit close, but there is nothing to suggest that the Saxtons had any interest in mining. Dunningley Hall was of modest dimensions; it would have been either stone built or of wood and plaster, raw materials for both being plentiful locally. Brick was not generally used in the area for some time to come.

The parish of Woodkirk was sparsely inhabited and contained several hamlets but no large settlement. The most important dates in the calendar were August 15th and September 8th when the woollen cloth fair was held at Lee Fair near the church, with visitors coming from considerable distances. Of more regular concern to local clothiers was the cloth market held twice weekly on the bridge over the river Aire at Leeds where manufacturers would walk or ride with their finished pieces. Wakefield however was the foremost centre of the woollen industry being considerably larger than either Leeds or Bradford, but restrictive practices at Wakefield gave advantages to the other centres, thus eventually losing Wakefield its former eminence. Until the dissolution of the monastic foundations a cell of black canons from the Augustinian Priory at Nostell, south-east of Wakefield, had served at Woodkirk and it is possible that some of the monks remained in the district in Saxton's early years.

The parish of Woodkirk was in the ancient Manor of Wakefield which reached from Normanton, near Pontefract, in the east, almost to Todmorden and the Lancashire border in the west. In the reign of Philip and Mary the manor was annexed and included in the Duchy of Lancaster with the reigning sovereign as Lord of the Manor. The chief centre of administration was Wakefield itself where the court baron was held every three weeks and the court leet twice yearly. Woodkirk was in the wapentake of Agbrigg which bounded the wapentake of Morley to the north. Roughly half the townships in these wapentakes were in the Manor of Wakefield, most of the remainder being in the Honour of Pontefract which was also part of the Duchy of Lancaster.

The local gentry and principal landowners of the neighbourhood included Sir Robert Savile, the illegitimate son of Sir Henry Savile of Thornhill, who commenced building Howley Hall to the west of Woodkirk church. Sir Robert died in 1585 and was succeeded by his son John who had a most illustrious career holding many high offices, was created Baron of Pontefract and became the first mayor of Leeds. Although it would appear likely that Saxton might have been employed by the Saviles of Howley none of their estate papers survive today to confirm this.

The Copleys of Batley Hall, Lords of the manor of Batley, were another influential family in the district, one of whom, Isabel, married Sir Robert Savile in 1563. It was on the marriage of Isabel's nephew, Alverey Copley, to her grand-daughter Elizabeth Savile in 1616, that Robert Saxton was employed to survey some land in connection with the marriage settlement.

Another local family of historical interest, but not of the gentry class, were the Fields of East Ardsley, certainly known to the Saxtons. John Field, who was a generation older than Christopher Saxton, can be named among the pioneers of science. In 1557 whilst residing in London, he published his Ephemeris, or astronomical tables, based on the theories of Copernicus, the first to be published in this country. John Field returned to Ardsley in the early 1560s where he lived until his death in 1587.[8] He was a friend of the celebrated John Dee who later employed Saxton to survey Manchester. Whether Field had any influence on Saxton's career cannot be said, but he could have been a useful contact.

There has been much speculation concerning Saxton's education and it has been claimed, without supportive evidence, that he was educated at the forerunner of Queen Elizabeth Grammar School, Wakefield, a school run by the clergy in the parish church of which there is plenty of evidence, making it an acceptable suggestion. J. W. Walker, for instance, in his history of the town says 'Christopher Saxton, born at Wakefield July 28, 1544, who passed on from this school to Cambridge, and was the great map-maker of every county in England in the sixteenth century, died 1596', without giving any reference to his source of information.[9] It has already been demonstrated that Saxton was alive well into the seventeenth century and came from Dewsbury parish, so on those points Walker is incorrect and the other facts are thus questionable. The date of birth does not appear in the first edition of Walker's history published in 1934, and according to his widow this information was supplied to Walker in the intervening years by Matthew Peacock, retired headmaster of the school and author of its history. Peacock had also claimed Saxton as a past pupil of the school. Peacock retired to Oxford but no corroborative evidence has come to light from there either.

It has likewise long been a supposition that Saxton was educated at Cambridge University, and in noting those said to have encouraged him in later years, this seems quite feasible. However none of the Cambridge colleges founded by the 1550s can trace any record of Saxton having studied there, although lack of evidence in itself is not conclusive as there are grounds for supposing that some students at that period went unrecorded. This was especially common in times of religious uncertainty, when some men went for the education but failed to graduate because they were not prepared to take the necessary oaths. The modern tradition of Saxton being a Cambridge man appears to date only from 1858 when an entry in Coopers' records of Cambridge men states that Saxton was 'A native of Wakefield, had his education at this university, but in what house or College or at what particular period we are unable to ascertain'.[10] In fact the only record of a man at university, bearing a name of close similarity at a time when spelling was unimportant, is of a Christopher Sexton who supplicated for a B.A. degree from Magdalen College, Oxford, 15 April 1586. In view of the late date it seems very unlikely that this was Saxton the surveyor.

David Marcombe has recently discovered a connection between Christopher Saxton and a man named John Rudd, who in all probability taught Saxton all he knew of the art and skill of cartography.[11] Rudd was vicar of Dewsbury from the end of 1554 to 1570, and rector of Thornhill to the south from 1558 to 1570 or 1578. He had been royal chaplain to Henry VIII and held a prebend at Durham Cathedral amongst other appointments. He had a consuming passion for cartography and as early as 1534 declared himself to have long been a student and teacher of the map-maker's art. Apparently Rudd had been engaged in making a 'platt' of England and in 1561 requested leave of absence from Durham for about two years. It was his intention to continue his cartographic work and to travel further 'for the setting forth thereof both fairer and more perfect and truer than it hath hitherto'. He wished to 'travel by his own sight to view and consider divers parts of our said realm by reason whereof he shall be forced for a certain time to be much absent from the said Church of Durham'. Approval of the Queen was obtained and the Chapter instructed to pay his emoluments as usual.

That Saxton was connected with Rudd is proven by a receipt signed by Saxton when he visited Durham in April 1570 to collect £8 6s 8d 'for thuse of my master Master Rudde for his quarter stypend dewe at thannuntcyacon last'.[12]

Saxton would have been ten or twelve years old on Rudd's arrival in Dewsbury and circumstantial evidence points strongly towards a long association between the two. Saxton may simply have progressed from being a bright school pupil, favoured by his master, to ultimately becoming his assistant. That Saxton was able to complete his national survey in so few years has always puzzled cartographic students, but it may be that in fact he had available and made use of the material Rudd had gathered on his travels, or indeed that he himself had accumulated if he had accompanied Rudd. Unfortunately none of Rudd's work is known so comparisons cannot be made. It is possible that the unnamed English friend, from whom Mercator obtained his information for his map of the British Isles, published in 1564, may have been Rudd.

The only contemporary surviving comment on Saxton is contained in a notebook now in the Bodleian Library at Oxford.[13] It was compiled by Dr John Favour, Fellow of New College, Oxford, and vicar of Halifax from 1593 until his death in 1623. The notebook was found in the library of Henry Foulis of Lincoln's Inn in 1669 and is mostly in Latin. There is one page devoted to Christopher Saxton, firstly giving details in Latin of his life and career, followed by one verse in Latin, and two in English, of what has become known as Saxton's 'epitaph'. The first paragraph translates:

'Christopher Saxton, born in Wakefield in the county of York (was) most skilled in geometry: having received letters from Queen Elizabeth dated 28 July in the 15th year of her reign, he travelled through the whole of England through towns and villages for nine continuous years with the utmost labour and industry and not only drew the counties separately and most carefully, but took the pains to have them engraved on bronze tablets. And then, to the everlasting memory of the undertaking, and in praise of his name, and to the advantage of the English state, he published and divulged it in the year of human salvation 1575.'

The verses follow:

'This humble earth bears Saxton's body
Who while he was living England scarcely recognised.
Now snatched away the earth bears his corpse

> But neither England nor sky can muffle his fame
> For although he is dead, he lives countrywide.
> Saxton alive wo^d England scarse mought hold,
> Lieth here interd in bass & cūtry mould.
> His body cladd in earth in land & skye his name
> Breaks forth, thoughe dead, hee lyves to cūtryes fame.
>
> 'If y^u for Saxton seeke, behold his grave,
> Yet h'is not here, he is in greater grace,
> The prince, ye nobles, gentils, learned haue
> Daygned him in court, in house, in study place.
> Ther seeke for Saxton's name, ther it is fownd,
> His earthly part is only in this grounde.
> His flesh in earth, his fame on earth is still,
> His soule at rest in heaven, attends Gods will.'

Dr Favour, the puritanical, anti-papist vicar had a reputation for composing humorous verses so they are quite probably his own work, although the possiblity remains that he copied them from elsewhere. Perhaps the most significant contribution to our knowledge of Saxton to be gained from the passage is that he received 'letters' or a commission from the Queen in 1573. This document has long since disappeared, but it is interesting to know, if it be true, that Saxton received royal patronage from that early date. If indeed Saxton took nine years over his surveying, he must have commenced about 1569, which seems unlikely in view of his connection with John Rudd in 1570. Favour ends his first paragraph by saying that the maps were published in 1575. In fact the atlas as a whole was published in 1579, but the individual maps became available, as completed, from 1574. By the time the first maps were engraved and ready for sale it may well have been 1575. The impression created by the passage is that it is correct in general terms but not in the particular.

The verses themselves tell us little. They are rather too long to have appeared on Saxton's tombstone and so cannot be connected with the parish church at Halifax as might have been implied had they been appropriate. Dr Favour was rarely in residence in Halifax, spending a great part of his time in York and elsewhere 'debating' with imprisoned Jesuits. In his travels he probably met Saxton, for the verses show his admiration for the man and the fact that he wrote them tends to confirm this view.

According to Fordham, Favour's notebook can be attributed to a period between the years 1603 and 1611, which latter date is further confirmation of Saxton's demise by that year.[14] The reference to Saxton having been born in Wakefield is possibly the source for later writers making the same statement. Favour may have meant the town, the parish, or the manor, of which we now know probably only the latter is factually correct and 'near Wakefield' would have been more appropriate.

Whatever the intimate details of Saxton's background, one thing is incontrovertible – he was chosen by Thomas Seckford of Seckford Hall, Woodbridge in Suffolk, to survey and map the counties of England and Wales. Thomas Seckford, as Saxton's patron, was virtually Saxton's employer and it was he who financed the undertaking. Seckford, son of Thomas Seckford, was born in 1515 and after graduating from Cambridge was admitted at Gray's Inn, London, in 1540, where he met William Cecil who was later created Lord Burghley. Thomas Seckford, as a competent lawyer, became Master of the Queen's Requests, Surveyor of the Court of Wards and Liveries, Steward of the Court of

Marshalsea, and Porter and Keeper of prisoners in the Marches of Wales. He became a wealthy man and considerable benefactor, founding and endowing almshouses in his native Woodbridge. His father died in 1575 when Thomas Seckford succeeded to his estates. He died in 1587.

Chapter 2

The National Survey

There is plentiful evidence that Christopher Saxton was by profession a surveyor. The record of extant manuscript plans is sufficient testimony to that, though surprisingly little is known of the man as an individual. More mysterious is the fact that his manuscript plans which we know of today date from c.1590, namely they follow his major cartographic achievement, and none earlier than 1574 appear to have survived. Saxton's association with a chart of Belfast Lough, 1569, is speculative and improbable.[1] One possibility must be that Saxton achieved sufficient reputation during his national survey to have been increasingly employed after it, and thus work subsequent to 1579 has survived by virtue of there being more of it in the first place. It remains perplexing, however, that a surveyor who was selected for a major national exercise should have so little testimony surviving from earlier years and that much of his later work was conducted in the comparative obscurity of his native Yorkshire.

(a) The Chronological Progress of Saxton's survey

His major achievement was indisputably the survey of English and Welsh counties begun about 1574 and completed by 1579. This was to be eventually issued during 1579 in atlas form, the first ever national atlas of England and Wales, though individual sheets were printed and distributed prior to that year. William Cecil, Lord Burghley, took great interest in the survey as it progressed. For this reason the maps were sent to him as each plate in succession was engraved, and represent the earliest proofs. It is by means of a study of these that the progress of the survey, to the extent that it is possible to recognise one, can most probably and accurately be assessed (Appendix 6). Even so, the probability is that Saxton availed himself of as much contemporary geographical information as possible, thus facilitating the speedy execution of his task (Chapter 4).

The Atlas, as it finally appeared, contained a printed index of the constituent county maps, but the Burghley proofs are found in an order corresponding to none of the printed indices, of which there are several variant forms. It can therefore be presumed that the Burghley proof maps were assembled without any foreknowledge as to their final order in Saxton's Atlas (though this, too, frequently varies from issue to issue), nor necessarily in chronological sequence. Indeed, the dates printed on the county sheets, assuming these to indicate the year of survey rather than engraving, bear out the last point. There is evidence, in fact, that Lord Burghley assembled his collection of maps as an atlas of his own, probably after he had acquired all Saxton's maps. These he rearranged, together with various other maps and notes, in an order which, with only a few exceptions, constituted the equivalent of a general atlas as understood today. The first map in Burghley's collection, for example, is Saxton's national map ('Anglia ...'), the last which Saxton actually produced during his work for the Atlas of England and Wales. There follows a number of county maps of south-western England, progressing from Cornwall towards Devon and Dorset. Inserted between Saxton's maps of Cornwall and Devon, however, there is an anonymous coloured manuscript showing the coast between Dartmouth and Weymouth, constituting an introduction to the two maps which follow it. The map of Dorset, in turn, is followed by manuscript maps of Falmouth and the Isle of Wight; the former completes the collection of maps on south-western England, and the latter is the first devoted to southern and south-eastern England. Notes frequently accompany maps where they serve to amplify the information contained on them, as witnessed by lists of Justices of the Peace. Despite this, there are several eccentricities of arrangement. For example, Burghley includes a map of north-western Europe, which excludes Britain and would appear irrelevant in a national atlas of this country; his map of Scotland is inserted after the plan of Jersey, whereas one might have expected it at least to precede this, or even to have immediately followed Saxton's maps of the northern counties of England; and the two county maps appertaining to Shropshire are inexplicably separated by two maps of Lancashire and one of Cheshire. Burghley was evidently unconcerned with the sequence in which Saxton surveyed the counties, despite his interest in the survey as a whole. His collection constitutes primarily an assemblage of maps in atlas format, probably effected after Saxton had finished his work. It is the nature and not the sequence of the Burghley proofs which is of relevance in attempting to discern the order in which Saxton produced his maps. Cartographic variations occur between some of the Burghley proofs and the maps as they were eventually issued by Saxton in atlas form, and it may therefore be supposed that where a Burghley proof was later subjected to additions or alterations, then that particular map was effected fairly early in the complete series.

There are five main cartographic features which assist in attempting to unravel the chronology of Saxton's survey:

(1) All Saxton's county maps except that of Northumberland bear dates (Table 1). Assuming that these indicate reasonably accurately the years during which the survey of various counties occurred, they are of immeasurable assistance in determining its progress.

(2) Most of the county maps contain no information relating to divisions smaller than the county itself, but five (Cornwall, Essex, Hertfordshire, Suffolk, and Norfolk) show the divisions into hundreds; on the first four the hundred names are engraved in geographical position on the face of the map, but for Norfolk the hundreds are lettered and a key to the letters and names is found in the top right-hand corner of the sheet. The Oxfordshire sheet makes no reference to hundreds, but indicates the number of parishes in each constituent county of the map in a panel in the top right-hand corner. It is to be observed from the dates engraved on these maps that two (Norfolk and Oxfordshire) were produced during 1574, that Suffolk and Essex belong to 1575, Cornwall to 1576, and Hertfordshire to 1577. Any reference to hundreds or parishes is clearly a feature of maps compiled comparatively early during the survey, though Hertfordshire is an exception to this rule.

(3) It has been observed that there are two states for each of the county maps of Norfolk and Northamptonshire.[2] The Burghley proofs of each are in State I; State II, which appears later in certain issues of Saxton's Atlas, is as follows:

    (a) Norfolk: the addition of three place-names.
    (b) Northamptonshire: the addition of two names and a ring-fence in Bedfordshire.

The fact that these additions (detailed in Chapter 3, Section b) were deemed necessary to the maps of these counties at a later date suggests that they belong to an earlier part of the survey.

(4) On the Burghley proof map of Cornwall the scale is engraved differently from that which appears in Saxton's Atlas as eventually published, and the word 'Regina' is omitted from the date panel. The alteration of the line-scale and addition of 'Regina' in the later issues suggests that Cornwall was surveyed comparatively early.

(5) Ten of the Burghley proofs occur in their final state, i.e. they include all those cartographic elements which, on maps preceding their final state, are occasionally omitted (Table 1). This suggests that by the time these particular counties were surveyed the general format of the survey had been well established, and that these counties were therefore among the last to be mapped.

Documentary evidence which assists in determining a chronological sequence for the survey is fourfold:

(1) On March 11th, 1574, Queen Elizabeth granted Saxton, in consideration of his expenses sustained 'in the survey of divers parts of England', a lease of lands at Grigston Manor in Suffolk (Appendix 7), Thomas Seckford's county. From this date onwards even proof copies of the county maps bore the Royal Arms, and almost all bore Seckford's motto (Table 1).

(2) On July 10th, 1576, Queen Elizabeth issued an order of assistance for Saxton's survey of Wales (Appendix 8). About this time Saxton must have been contemplating his Welsh survey, although this did not, in the event, begin until 1577.

(3) Sometime during 1576, Thomas Seckford altered his motto on the maps from 'Pestis patriae pigrities' to 'Industria naturam ornat'. This new motto was not used by the Seckford family, whose motto was 'Win't and Wear't'. Lynam[3] has suggested that 'Pestis patriae pigrities' ('Sloth is the curse of the fatherland') referred to the energy of the author, and 'Industria naturam ornat' ('Industry adorns nature') to the artistry of the engravers. Early maps bear the older motto, and later ones the more recent.

(4) On July 20th, 1577, Queen Elizabeth granted Saxton a licence for the exclusive publication of his maps during a period of ten years (Appendix 9). Lynam[4] attempts to explain the granting of the licence in 1577 rather than 1574, when the first maps were engraved, or in 1579, by which time the survey was completed, since its granting in 1577 curtailed Saxton's monopoly of sale by two years. It is possible that the publication in 1577 of Holinshed's *Chronicles* stimulated a demand, or was expected to, for single maps. This book, originally conceived of as Wolfe's *Cosmographie*, was designed to have contained Wolfe's maps, but the *Chronicles* appeared without them. Holinshed and William Harrison (who wrote the *Description of Britain* and translated the *Description of Scotland* for the *Chronicles*) would have therefore encouraged Seckford's Atlas, because it would illustrate and help sell their own book. In his dedication of the work to Lord Burghley, Holinshed acknowledged the assistance he had received from 'Maister Sackford's cardes', and Harrison thanked Seckford for the loan of his 'platformes', dedicating the *Description of Scotland* to him ('Having by your singular curtesie receyved great help in my description of the rivers and streams of Britain and by conference of my travaile with the platformes of those few shires of England which are by your infinite charges alreadie finished'). Saxton is not mentioned, but until this date his name had not appeared on the maps. Whatever the reason, the granting of this licence stimulated the production in 1577 of more maps than during any other year of the complete survey. Further, the sale of the maps presumably started during this year, and Saxton's name thus began to appear on his maps for the first time (Table 1).

Further evidence relating to the chronological progress of the survey lies in comparatively recent research, of which however very little has been undertaken to this end. It can be reasonably presumed that, where possible, Saxton proceeded from one county to the next, following some convenient itinerary. This supposition is in many cases the only hypothesis one can use in attempting to determine a chronological sequence of events. Manley,[5] on the basis of evidence from the Lancashire and Yorkshire maps, has suggested that Saxton surveyed Yorkshire prior to Lancashire, but against this proposal there have to be set a number of other features on the Burghley proofs which will be discussed later.

(b) A Suggested Chronological Sequence

1574 Maps The earliest dated maps are those of Norfolk and Oxfordshire; in both there are forceful indications that a uniform system of cartographic compilation for the entire series of maps had not yet been worked out, and that these two sheets were perhaps to some extent experimental. The map of Norfolk is divided into hundreds (as also are the maps of Cornwall, Essex, Hertfordshire, and Suffolk), but it is the only map so divided which uses a system of lettering for each hundred on the map with an accompanying explanatory legend of hundred-names (in the top right-hand corner). Further, the map was evidently judged later to be somewhat incomplete, because three extra place-names were added at an unknown date; (these features have already been observed). The map of Oxfordshire does not show hundreds, but rather gives the number of parishes in each county shown on the map, again in the top right-hand corner. The Norfolk map bears neither the Royal Arms nor Seckford's motto; the Royal Arms were not added until 1577, when the map appears in its final state. Neither does the map of Oxfordshire contain Seckford's motto, but the Royal Arms do appear, evidently inserted late in the top left-hand corner of the map.

It is possible that Norfolk was the first of the counties to be surveyed by Saxton, and that the Burghley proof was completed quite rapidly in 1574; too early, in fact, for the addition of the Royal Arms and Seckford's motto. The rapidity of the survey may be surmised from the omission of three place-names added subsequently. The map of Oxfordshire, however, might have been produced somewhat later, permitting the addition, albeit as an afterthought, of at least the Royal Arms (this would have received priority over Seckford's motto, which was still omitted and not added until a later date). It is therefore reasonable to suggest that Norfolk was the first county to be surveyed, and that those occurring on the Oxfordshire map came next.

1575 Maps six maps are dated 1575 (Table 1), and all but one (Suffolk) are of counties in southern England. Among those sheets engraved with the date 1576, the following year, Cornwall is the only one on the south coast. It is therefore tempting to suggest that in 1575 Saxton's main task was a survey of southern England, working his way systematically westwards towards Cornwall.

Suffolk, however, lies outside this immediate area, and there are reasons for presuming that it was the first of the 1575 maps to be finished. It would complete Saxton's survey of East Anglia, already begun in Norfolk. Further, Queen Elizabeth's grant to Saxton during 1574 of a lease of lands at Grigston Manor in Suffolk, together with the fact that this was his patron's home county, might have encouraged him to map it as soon as possible. It has already been remarked that following the Queen's grant to Saxton, even proof copies of his maps bore the Royal Arms and most bore Seckford's motto. On the Burghley map of Suffolk, however, Seckford's motto does not appear (it was inserted later, replacing the name of the engraver, Lenaert Terwoort). The map resembles in this respect those of Norfolk and Oxfordshire, suggesting that it succeeded them without any significant interval and that it was completed too soon for the insertion of Seckford's motto on the proof copy. The supposition is given extra weight when it is observed that hundreds are also shown, just as they had been for Norfolk during the previous year.

Presuming that Saxton's survey in 1575 took him progressively further westwards through southern England, the county maps would assume the following chronological sequence: Kent; Hampshire; Dorset; and Devon. If one inserts Somerset between Dorset and Devon, the theory has the added attraction of suggesting that Saxton handed his surveys to two engravers alternately for engraving, i.e. Kent to Hogenberg, Hampshire to Terwoort, Dorset (with no engraver's name, but in Hogenberg's style), Somerset to Terwoort, and Devon to Hogenberg. It is also of interest to note that Cornwall, more than likely the next county to be surveyed, was engraved by Terwoort. There is clearly some attraction in such a theory, but no further evidence exists to support it. Burghley's map of

Somerset has a blank title-panel with the word 'Summersetshire' added in manuscript, and it lacks Seckford's motto. The map was clearly incomplete when Burghley received it, and this may indicate a hurried effort by Saxton or Terwoort, though there is no readily perceivable reason why this should have been necessary.

It appears likely, therefore, that in 1575 Saxton first surveyed Suffolk, and that he then began his survey of southern England, progressing systematically westwards from Kent; but one cannot be definite in assigning a chronological sequence to the southern survey, and the mystery of an incomplete map of Somerset remains.

1576 Maps Eight maps bear the date 1576, and several significant observations may be made regarding them, viz:

(1) Five bear Seckford's old motto, 'Pestis patriae pigrities', i.e. Cornwall, Durham, Essex, Northamptonshire, and Warwickshire, while the remainder have the new motto ('Industria naturam ornat'). One may therefore assume that these five county maps were completed prior to the change in motto, i.e. that they preceded those with the new motto. Two, however (Warwickshire and Durham), also carry Seckford's new motto, and therefore probably followed those with only the earlier motto on the map, the change in motto no doubt having occurred during the compilation of these two maps.

(2) The maps of Cornwall and Essex display the earlier characteristic of showing hundreds, reinforcing the suspicion that they belong to the earlier part of 1576. On the map of Cornwall, too, the scale is differently engraved to that in its final state, and the word 'Regina' is omitted from the date panel. The map of Northamptonshire, in view of its occurrence in State I, also appears to fall early in the sequence.

(3) Saxton had already been working in south-western England at the close of 1575, and it is reasonable to assume that he proceeded westwards into Cornwall to complete his survey of the south coast counties early in 1576.

(4) The maps of the counties of Lincolnshire (including Nottinghamshire), Westmorland (with Cumberland), and Wiltshire all bear Seckford's new motto, 'Industria naturam ornat', and they lack those cartographic features characteristic of the earlier maps.

It is thus at least possible to propose a sequence for 1576. Saxton first completed his survey of Cornwall, subsequently moved into Essex to finish his survey of eastern England, and then proceeded to the east Midlands, where he successively mapped those counties shown on the maps of Northamptonshire and Warwickshire. It is impossible, however, to be precise regarding the chronological sequence of these last three maps. Saxton might then have travelled north to commence his survey of northern England, beginning with Durham and also covering adjacent Westmorland/Cumberland. Subsequently he moved southwards to begin his survey of the west Midlands and/or Wales (having received an order of assistance for the latter survey in 1576). The large area of Yorkshire and Lancashire was omitted for the time being, but it seems he surveyed and mapped Lincolnshire/Nottinghamshire and Wiltshire.

It is unfortunate that the details we have to assist us do not permit a more accurately assessed chronological sequence for 1576, but the conclusions reached are not unreasonable. The only undated map is that of Northumberland. It is included at this point for the following reasons:

(1) Like all the maps so far examined it contains no reference to Saxton (which does not appear on any county maps in the Burghley collection until 1577).

(2) It bears Seckford's new motto, so clearly is later than the map of Durham, belonging either to the Westmorland/Cumberland group or to some later period.

(3) It forms a geographically undated 'island' in the same region as Durham and

Westmorland/Cumberland, both belonging to 1576 and both engraved by Ryther (although there is no engraver's name on the map of Northumberland).

It would seem more than likely, therefore, that Northumberland was mapped during the latter part of 1576, following Durham, but in a chronologically uncertain position with respect to the survey of Westmorland/Cumberland.

1577 Maps It is impossible to be precise with regard to the progress of the survey during 1577. Twelve of Saxton's county maps bear this date, the largest number issued during any one year. Possibly this was the result of the licence granted him in 1577 by Elizabeth for the exclusive publication of his maps; during this year his name began to appear on the maps for the first time. It has been noted that towards the close of 1576 he seems to have been moving southwards, possibly with a view to beginning his survey of Wales. It appears that in the event he decided to complete first the outstanding English counties (though he did survey Denbighshire and Flintshire). But the determination of a sequence of progression is difficult. The only positive evidence which can be used relates to the engraving of Saxton's name on his later maps (a process which had been effected on all thirty-five maps in the Atlas by the time it was issued in 1579; see Chapter 3). Some of the Burghley proof maps do not possess such an acknowledgement, i.e. Cheshire, Derbyshire, Herefordshire, Lancashire, Shropshire, Staffordshire, and Worcestershire. As it was in July of this year that Saxton received a licence for the exclusive publication of his maps, it may be assumed that those maps not bearing his name belong to the earlier part of the year, and that the others came later, i.e. Denbighshire/Flintshire, Gloucestershire, Hertfordshire, Monmouthshire, and Yorkshire. This view is given added weight when it is observed that the latter maps in the Burghley collection exist in their final state, i.e. that no subsequent additions or amendments were considered necessary.

From these observations it is only possible to divide the 1577 maps into two groups, the first earlier than the second:

(a) Cheshire, Derbyshire, Herefordshire, Lancashire, Shropshire, Staffordshire, and Worcestershire.

(b) Denbighshire/Flintshire, Gloucestershire, Hertfordshire, Monmouthshire, and Yorkshire.

It can be suggested, therefore, that Saxton moved from Wiltshire (already supposed to be his last position in 1576) towards Wales, for which he had in 1576 received an order of assistance. In view of his receiving, in 1577, a licence to publish his maps, however, he seems to have preferred to press on with the completion of the English survey; starting in Herefordshire, moving northwards along the Welsh border through Shropshire and Cheshire into Lancashire, and then back southwards through Derbyshire, Staffordshire and Worcestershire into, during the latter part of 1577, Gloucestershire and Monmouthshire. He may then have decided to complete Yorkshire, the largest English county, prior to beginning his survey of Wales in Denbighshire/Flintshire. Such a sequence is not illogical, but is clearly rather generalised and tentative.

Two particular problems arise in assigning a chronological sequence to the 1577 maps. First, Manley[8] has suggested that Saxton might have surveyed Yorkshire prior to Lancashire, a conclusion opposite to that suggested by the occurrence or otherwise of Saxton's name on the Burghley proofs. Second, the map of Hertfordshire appears to be an anachronism in the sequence. Saxton's name is inserted on the map, but the scroll containing it is squeezed in, suggesting that it was added very shortly after Saxton received his licence. It is possible to suggest, though most speculatively, that Saxton rapidly surveyed Hertfordshire in between his surveys of Worcestershire and Gloucestershire, and entrusted it to a London engraver who was not otherwise engaged on the survey at all, and whose style bore little resemblance to the rest of the series. The issue of Saxton's licence may have followed very soon upon the completion of the map itself, and the acknowledgement could have been inserted as a rather cumbersome addition on Saxton's instructions.

1578 Maps The five maps bearing the date 1578 all relate to Welsh counties. They all acknowledge Saxton, all bear Seckford's new motto, and are all in their final state. There is, therefore, little to assist us in deciding upon a chronological sequence. If, as it appears, Saxton began his Welsh survey in Denbighshire/Flintshire in 1577, it is reasonable to suppose that he proceeded through the Welsh counties generally from north to south, i.e. Anglesey/Caernarvonshire, Montgomeryshire/Merionethshire, Radnorshire/Breconshire/Cardiganshire/Carmarthenshire, with either Pembrokeshire or Glamorganshire last. It is impossible to be more definite than this, nor can it be said in what precise order he surveyed the counties within each of the map groupings which account for most of the Welsh survey.

Conclusion (Table 2) It would seem that Saxton began his survey somewhat experimentally in Norfolk and the south Midlands. Subsequently he surveyed Suffolk and, progressing westwards through the area, the south coast counties of England. From here he moved into Essex and the east Midlands, and then into the extreme north of England. In 1577 he completed his English survey and began the survey of Wales which he completed in 1578. The evidence that can be used in determining a chronological sequence becomes increasingly difficult to interpret as the years proceed. There is a measure of certainty until early in 1576, but the maps for most of this and subsequent years are extremely difficult to place in sequence.

TABLE 1

Saxton's county maps in the Burghley Atlas

(Arranged chronologically according to engraved dates, and alphabetically for each year)

Columns: (i) Map subject.
        (ii) Seckford's motto (State a or b) (Note 1).
        (iii) Acknowledgement of Saxton's part in the survey, and spelling of his Christian name (Note 2).
        (iv) Map state and associated details.
        (v) Engraver.
        (vi) Notes relevant to chronological placing.

All the county maps include Seckford's arms, and all except Norfolk the Royal Arms. The map of 'Anglia...' (not included in the table) was engraved by Ryther in 1579, and incorporates the Royal Arms, Seckford's arms and motto (State b), and an acknowledgement of Saxton ('Christophorus Saxton descripsit').

Year	(i)	(ii)	(iii)	(iv)	(v)	(vi)
1574	Norfolk			Earliest state. (Place-names added later.)	de Hooghe	Hundreds shown; key to names in top right-hand corner of map. No Royal Arms nor Seckford's motto.
	Oxon./Bucks./Berks.				? Rutlinger (Note 3)	Number of parishes in each county given in top right-hand corner of map. No Seckford motto. Royal Arms evidently added late in top left-hand corner.

Year	(i)	(ii)	(iii)	(iv)	(v)	(vi)
1575	Devon	a			Hogenberg	
	Dorset	a				
	Hampshire	a			Terwoort	
	Kent/Sussex/ Surrey/Mddx.				Hogenberg	
	Somerset				Terwoort	Blank title-panel with word 'Summersetshire' added in ms. No Seckford motto (Note 4).
	Suffolk				Terwoort (Note 5)	Hundreds shown; names in place on face of map. Seckford's motto omitted (Note 5).
1576	Cornwall	a			Terwoort	Hundreds shown; names in place on face of map. 'Regina' omitted from date panel. Early version of line-scale.
	Durham	a/b (Note 6)			Ryther	
	Essex	a				Hundreds shown; names in place on face of map.
	Lincs./Notts.	b			Hogenberg	
	Northants/Beds./ Cambridge/Hunts./ Rutland	a		Earliest state. (Place-names and ringfence added later.)		
	Warwick./Leics.	a/b (Note 7)			Terwoort	
	Westmor./Cumb.	b			Ryther	
	Wilts.	b			Hogenberg	
1577	Cheshire	b			Scatter	
	Denbigh/Flint	b	Christoferus	Final state	Hogenberg	
	Derby	b				
	Gloucs.	b	Christoferus	Final state	Ryther	
	Hereford	b			Hogenberg	

Year	(i)	(ii)	(iii)	(iv)	(v)	(vi)
	Herts.	b	Christoferus	Final state	Reynolds	Differs in many respects from other maps in the Atlas. Hundreds and names in position on face of map. Some names clearly touched up in ink by Burghley.
	Lancs.	b			Hogenberg	
	Monmouth	b	Christophorus	Final state		
	Salop	b			Hogenberg	
	Staffs.	b			Scatter	
	Worcs.	b				
	Yorks.	b	Christoferus	Final state	Ryther	
1578	Anglesey/Caerns.	b	Christophorus	Final state		
	Glam.	b	Christophorus	Final state		
	Mont./Merioneth	b	Christoferus	Final state	Hogenberg	
	Pembs.	b	Christoferus	Final state		
	Radnor/Brecon/ Cards./Carms.	b	Christoferus	Final state	? Scatter (Note 8)	
No date	Northum.	b				

Notes
1. Seckford's motto: (a) 'Pestis patriae pigrities'; (b) 'Industria naturam ornat'.
2. Saxton acknowledgement: 'Christoferus Saxton descripsit', or 'Christophorus Saxton descripsit'.
3. R. A. Skelton[6] says that the Burghley proof shows an erased inscription, 'Johannes Rutlinger (sculpsit?)'. While a space does exist at the foot of the title-cartouche, there is no clear evidence of such words occurring.
4. Seckford's arms appear to have been added rather cumbersomely in the bottom right-hand corner of the map. The scroll which ultimately was to contain Seckford's motto ('Pestis...') is shown, but is left blank.
5. Seckford's motto ('Pestis...') was eventually to replace the name of Lenaert Terwoort, which occupies a small cartouche beneath Seckford's arms.
6. 'Pestis patriae pigrities' and 'Industria naturam ornat' both occur, below and above Seckford's arms respectively.
7. 'Pestis patriae pigrities' occurs in a scroll below Seckford's arms. 'Industria naturam ornat' is added above Seckford's arms.
8. A. N. Hind[7] says that the Burghley proof indicates the map to have been engraved by Francis Scatter. There appears, however, to be no basis for such an attribution.

TABLE 2

A suggested chronological sequence for Saxton's county surveys

Dates in the first column indicate only the assumed chronological sequence of maps. A more detailed, though less certain sequence, is indicated by numbers listed beneath various dates.

Year	Map	Notes
1574		
1	Norfolk	
2	Oxon./Bucks./Berks.	
1575		
3	Suffolk	
4	Kent/Sussex/Surrey/Middlesex	
5	Hampshire	
6	Dorset	
7	Devon	
8	Somerset	Possibly precedes Devon.
1576		
9	Cornwall	
10	(a) Essex (b) Northants./Beds./Cantab./Hunts./Rutland	Chronological sequence with respect to one another uncertain.
11	Warwickshire/Leicestershire	
12	(a) Durham (b) Northumberland (c) Westmorland/Cumberland	Chronological sequence with respect to one another uncertain.
13	(a) Lincs./Nottinghamshire (b) Wiltshire	Chronological sequence with respect to each other and to those maps in the table above is uncertain, and positioning is based solely on a logical geographical itinerary.
1577		
14	(a) Herefordshire (b) Shropshire (c) Cheshire (d) Lancashire (e) Derbyshire (f) Staffordshire (g) Worcestershire	Chronological sequence with respect to each other is based solely on a logical geographical itinerary. Manley[9] might challenge the inclusion of Lancashire in this section.
15	Hertfordshire	Chronological position difficult to determine with any degree of precision.
16	(a) Gloucestershire (b) Monmouthshire	Probably executed in succession to one another, but chronological sequence with respect to one another or to those maps in the table below is uncertain.
17	Yorkshire	Manley[10] would probably place this county earlier in the sequence.
18	Denbighshire/Flintshire	

Year	Map		Notes
1578			
19	(a)	Anglesey/Caernarvon.	Chronological sequence based on a logical geographical itinerary.
	(b)	Mont./Merioneth	
	(c)	Radnor/Breck./Cards./Carms.	
20	(a)	Pembrokeshire	Probably last maps in sequence, but chronological position with respect to one another uncertain.
	(b)	Glamorganshire	

Chapter 3

Saxton's Atlas of England and Wales, 1579, and wall-map of 1583

Saxton's Atlas of England and Wales appeared in 1579. It comprised a map of England and Wales ('Anglia...'), the county maps, and several introductory leaves. The essential contrast between Saxton's assemblage of maps and that of Lord Burghley, to whom the maps had already been presented in proof form (see Chapter 2), was that the former constituted a true atlas, while the latter was merely a collection of sheets rather loosely, though not illogically, brought together. Chubb[1] somewhat confusingly lists the Burghley collection as one among several copies of Saxton's Atlas (Entry IV). It was nothing of the kind, differing from the latter in several ways. Chubb describes it as 'another impression' of Saxton's Atlas, wanting the frontispiece, the index to the maps, and the table of coats-of-arms. Yet it is the very nature of the omissions, both of introductory leaves and of cartographic design features, which indicates that Burghley was not, as Chubb seems to imply, aessmbling a definitive atlas from Saxton's plates. By the same token, the comparatively modest number of surviving copies of the Atlas (some in original sixteenth century binding), and the uniform format of the maps, demonstrate that Saxton's publication had throughout been conceived of as an atlas.

(a) Preliminary Leaves
Saxton's Atlas includes usually, though not invariably, three introductory pages. These constitute, quite clearly, the opening section of the Atlas, and are missing from Burghley's collection. They consist of:
   (a) a frontispiece;
   (b) an index to the maps in the Atlas;
   (c) a list of cities and towns, etc., which is usually accompanied by a preceding table of coats-of-arms.
Following these introductory pages the Atlas consists entirely of Saxton's cartographic work, though occasionally maps deriving from other sources have been interpolated and appended into various copies.

The engraved frontispiece is printed on the single side of a leaf of the Atlas, normally the right-hand side. It measures 378 × 238 millimetres, and has been variously attributed to R. Hogenberg and A. Ryther, but is more probably by the former. It depicts Queen Elizabeth, holding a sceptre and crown, enthroned under a canopy surmounted by the Royal Arms and two cherubs holding a wreath of laurels in each hand. Elizabeth was the patron of geography and astronomy, the latter being represented by bearded male figures on either side, one bearing compasses and a globe and the other an armillary sphere. At the bottom of the page is a Latin verse of six lines dated 1579, flanked on the right by an astronomer taking observations and on the left by a cartographer using a graver. The frontispiece is usually, but not invariably, hand-coloured, does not appear in all copies of the Atlas, and occurs in two states. State I (Plate 1) shows the Queen's dress, which has elaborate and jewelled ornamentation, lying in a hard horizontal line across her lap. Very

Frontispiece to Saxton's Atlas, 1579. Detail of Queen Elizabeth enthroned (State I). *The Queen's dress has elaborate and jewelled ornamentation, and lies in a hard horizontal line across her lap.*

PLATE I

COMITATVVM SINGVLORVM ISTO VOLV-
mine descriptorum index, ordinem quo cuiusque inueniatnr carta gra-
phica recte demonstrans.

	Anglia.
1	Herefordia.
2	Salopia.
3	Staffordia.
4	Wigornia.
5	Oxonium, Buckinghamia, & Berceria.
6	Hartfordia.
7	Northamptonia, Bedfordia, Cantabrigia, Huntingdonia: & Rutlandia,
8	Warwicensis, & Lecestria.
9	Derbia.
10	Cestria.
11	Lancastria.
12	Westmorlandia, & Cumberlandia.
13	Northumbria.
14	Dunelmensis Episcopatus.
15	Eboracensis: Comitatus.
16	Lincolnia & Nottinghamia.
17	Norfolcia.
18	Suffolcia.
19	Essexia.
20	Cantium, Southsexia, Surria & Middlesexia.
21	Southamptonia.
22	Dorcestria.
23	Wiltonia.
24	Deuonia.
25	Cornubia.
26	Somersetensis.
27	Glocestria.
28	Monumetha.
29	Glamorgana.
30	Penbrok.
31	Radnor, Brecknok, Cardigan, & Caermarden.
32	Montgomeri, ac Merionidh.
33	Anglesei, & Caernaruan,
34	Denbigh, ac Flnt.

Index to Saxton's Atlas, 1579. Setting A, Variant II.

COMITATVVM SINGVLORVM ISTO VOLVMINE
descriptorum index, ordinem quo cuiusque inueniatur
carta graphica, recte demonstrans.

✠	Anglia.
1	Herefordia.
2	Salopia.
3	Staffordia.
4	Wigornia.
5	Oxonium, Buckinghamia, & Berceria.
6	Hartfordia.
7	Northamptonia, Bedfordia, Cantabrigia, Huntingdonia, & Rutlandia.
8	Warwicum, & Lecestria.
9	Derbia.
10	Cestria.
11	Lancastria.
12	Westmorlandia & Cumberlandia.
13	Northumbria.
14	Dunelmensis Episcopatus.
15	Eboracum.
16	Lincolnia & Nottinghamia.
17	Norfolcia.
18	Suffolcia.
19	Essexia.
20	Cantium, Southsexia, Surria, & Middlesexia.
21	Southamptonia.
22	Dorcestria.
23	Wiltonia.
24	Deuonia.
25	Cornubia.
26	Somersetus.
27	Glocestria.
28	Monumetha.
29	Glamorgana.
30	Penbrok.
31	Radnor, Brecknok, Cardigan, & Caermarden.
32	Montgomeri, ac Merionidh.
33	Anglesei, & Caernaruan.
34	Denbigh, ac Flint.

Anno Domini. 1579.

Index to Saxton's Atlas, 1579. Setting B.

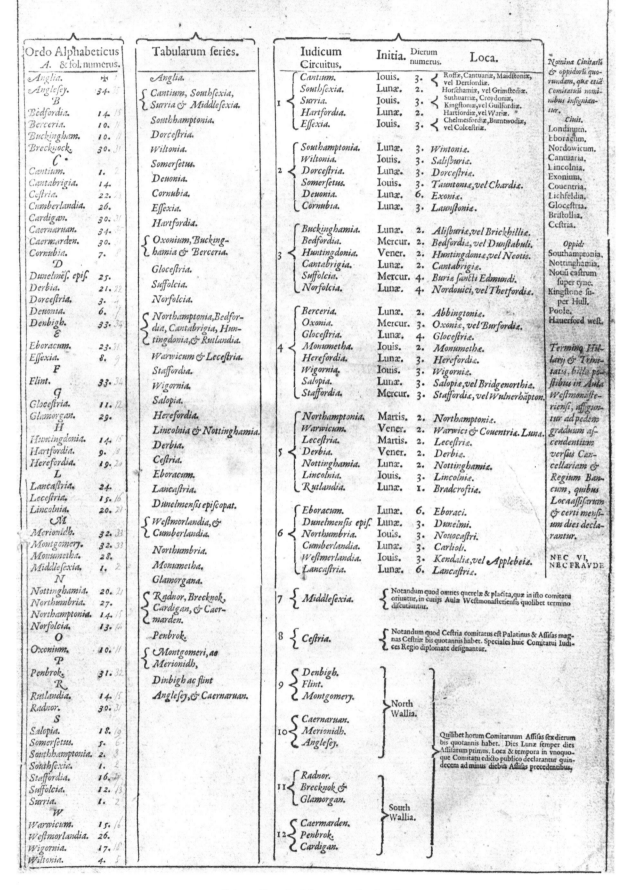

Index to Saxton's Atlas, 1579. Setting D.

few impressions are known to exist in this state, and only one certainly (Appendix 10, Copy No.45) and one probably (Appendix 10, Copy No.36) belong to the copies of the Atlas in which they occur; additionally there exist two single sheets unaccompanied by maps. State II (Frontispiece) results from the re-engraving of the Queen's forearms and dress, with the jewellery much simplified and the drapery falling in natural folds from her knees. Hind[2] suggests that the alteration might have been due to either the change in costume from a stiffer fashion to one nearer that of Hilliard's Great Seal of 1586, or to the Queen's personal suggestions for the modifications. One of the earliest known copies of the Atlas (Appendix 10, Copy No.5) has no frontispiece, but its position is occupied by a blank leaf of the original paper. It is possible that when the maps in this volume of the Atlas were printed either the frontispiece had not yet been engraved, or it was undergoing revision and was not available for printing. Certainly no date for the change from State I to State II can be definitely asserted; but in view of the rarity of State I, it may be suggested that the change occurred very early, perhaps even in 1579.

The index of maps was the only page of the Atlas to be printed from movable type. It was printed on one page, usually the right of a single leaf, and occurs in four different type-settings. Setting A (terminology used here and subsequently follows that established by Skelton[3]) is in roman, the first line of the three-line heading being in capitals and the last in italics, with ruled lines surrounding the type-page. Beneath the heading occurs a list of the maps in the Atlas, beginning with 'Anglia ...' (unnumbered, but marked by a Maltese cross) and continuing with the county maps (numbered 1–34 in a single column). The index is undated. Only four copies of the Atlas are known with this setting, and two variant forms have been recognised:

    Variant I   (e.g. Appendix 10, Copy No.36). Five misprints, i.e. 'Eboraceusis', 'Cantinm', 'Mersonidh', 'Flnt', and 'inueniatnr'.

    Variant II   (e.g. Appendix 10, Copy No.7). Three of these misprints have been corrected, i.e. 'Eboracensis', 'Cantium', and 'Merionidh' (Plate 2).

These two variant forms suggest that the type-compositor was working from a manuscript copy, and for this reason Setting A is normally given chronological priority over the others.

Setting B (Plate 3) is known to occur in only four copies of the Atlas (Appendix 10). The heading, in roman, again occurs in three lines, the first in capitals and the last in italics. But the type has been reset to provide alternative line-endings, and the page does not possess ruled border lines. The list of maps is the same as in Setting A (suggesting that this is also an early state) but the titles given to some of the maps vary slightly (i.e. numbers 8, 15, 26, and 34). The date 1579 occurs at the foot of the list, compounding the impression that this was an early version of the index.

Setting C is known to occur in only one copy of the Atlas (Appendix 10, Copy No.44). It differs from those preceding it with regard to the following features:

    (a) the type-setting of the title and its spelling;

    (b) the list of maps. This is still set as a single column numbered 1–34, but the titles are in italics, and the order of the maps varies. 'Anglia ...' is omitted, and the sequence of the maps is more akin to that in Setting D, but with the seven maps of the Welsh counties interpolated as maps 18–24. The first map in the list, as in Setting D, is 'Cantium ...', but the last is 'Northumbria'.

The index has an affinity with Setting A in being undated, but the similarity of the map sequence with that in Setting D, and the fact that both Settings C and D use an identical typeface, suggests that this setting immediately preceded the latter.

Setting D (Plate 4) is the most common setting of the index in extant copies of Saxton's Atlas. Its heading occurs in four lines, and this is followed by three columns; the first lists

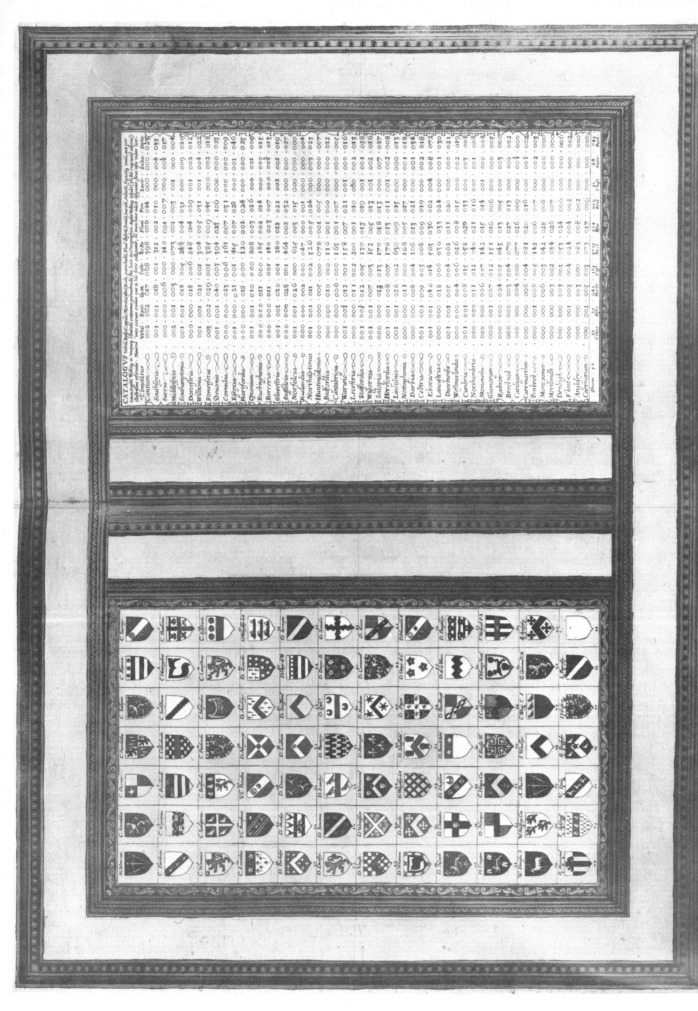

Saxton's Atlas, 1579. Plate of coat-of-arms and table of cities. The two leaves are printed from a single plate; in some copies the right-hand page is blank, representing either an earlier state or possibly the result of masking or mutilation.

PLATE 5

Saxton's Atlas, 1579. Map of 'Anglia . . .'. State II, showing the addition of a graduation for latitude and longitude inside the original border.

PLATE 6

the counties in alphabetical order, with number-references to the maps (though 'Anglia...' carries a Maltese cross); the second column represents a list of the thirty-five maps in their final order (which varies from that in preceding settings, though is similar to that in Setting C), without numbers; and the third column is a table of judges' circuits and assizes. It is likely that this setting is dated c.1590, because it usually occurs in copies of the Atlas possessing the table of coats-of-arms, which can be dated as later than September, 1589, and probably prior to 1591 (see subsequent paragraphs), and because one copy of the Atlas containing this setting (Appendix 10, Copy No.4) is bound up with, and printed on the same paper as, Ryther's plates of the Spanish Armada engraved to accompany P. Ubaldini's *Discourse*, published by Ryther in 1590 (both Atlas and Armada Plates reveal a watermark of crossed arrows). Ryther's plates, on paper carrying the watermark of a bunch of grapes, occur too in a collection of old maps in the possession of the Royal Geographical Society (264.g.), amongst which are also found thirty individual county maps by Saxton, similarly on paper bearing the watermark of a bunch of grapes.

The coats-of-arms and table of cities (Plate 5) is engraved on a single plate, occupying one leaf of the Atlas (388×522 millimetres within its border). The left-hand page consists of eighty-four coats-of-arms, one of which, in the bottom right-hand corner, is blank. The first sixty-five consist of the coats-of-arms of peers, while the remainder are those of officers of state. The right-hand page (left blank in some copies, representing either an earlier state or possibly the product of masking or mutilation, e.g. Appendix 10, Copy No.10) is a table of towns, etc., in each county. At the foot of the plate, outside the border of the table, is a small 'Scala miliarium', giving the relative lengths of the long, middle, and short miles. The dates of tenure of office by the owners of certain coats-of-arms, for example Sir Christopher Hatton (Coat-of-arms, No.66) noted as Chancellor, an office to which he was appointed in 1587 and held until his death in 1591, and Sir Thomas Heneage (Coat-of-arms, No.69) noted as Vice-Chancellor, an office he assumed in 1589, show that this plate cannot have been issued before 1589 or after 1591, except in the latter case through oversight.[4]

(b) Salient Cartographic Characteristics

The maps in Saxton's Atlas differ in several instances from the same maps as they were initially seen by Burghley in proof form. Variations are as follows:

(1) the addition of the Royal Arms of Queen Elizabeth to the map of Norfolk;

(2) the addition of Seckford's motto to the maps of Kent, Norfolk, Oxfordshire, Somerset, and Suffolk (in the last instance replacing the name of the map's engraver, i.e. Lenaert Terwoort);

(3) the addition of a Latin title to the map of Somerset;

(4) the re-engraving of the line-scale on the map of Cornwall, and the addition of the word 'Regina' to the dedication;

(5) the addition of Saxton's name to all those maps which, during and prior to 1577, had originally been engraved without it.

Only ten maps remained unmodified from the proof form in which Lord Burghley received them, viz: Anglesey, Denbighshire, Glamorganshire, Hertfordshire, Monmouthshire, Montgomeryshire, Pembrokeshire, Radnorshire, Gloucestershire, and Yorkshire (Table 1).

Three maps were to undergo revision after their initial incorporation into Saxton's Atlas, so that each occurs in two states. These are the maps of 'Anglia...', Norfolk, and Northamptonshire (Table 1). State II of the map of 'Anglia...' is indicated by the appearance of a graduation for longitude and latitude in the border, a feature wanting in State I (Plate 6). The map of Norfolk reveals in State II the addition of three place-names ('E.

Somerton', 'Heringbye', and 'Oxned'), while State II of the map of Northamptonshire has two names ('Taternall' and 'Merston') and a ring-fence and two trees (at 'Woodende') inserted in Bedfordshire. Future research may yet reveal further states of Saxton maps. An early proof of the map of Hampshire (Bodleian Library, Gough Maps Hants 16) precedes even the copy in the Burghley proof collection, the latter being the second known state of this map with many additions of trees and lettering; a third and fourth state are also recognisable.[5]

(c) Cartobibliographical Significance

Extant copies of Saxton's Atlas did not emerge as a single issue; some appeared prior to others. The determination of a chronological sequence for these issues is difficult, owing to a lack of records of its printing and publication. Much interest has therefore focussed on the paper used, but almost all the watermarks found in early issues (crossed arrows or a bunch of grapes, both in variants) were in use throughout the late sixteenth and early seventeenth centuries, and are of little use in dating the various issues or assigning any of them a chronological priority. It might be possible to determine some sequence through a more precise comparative study of the variant watermarks,[6] or by determining the sharpness or wear of the plates in known copies,[7] but this is a difficult and formidable task, producing in the event inconclusive results. Such a situation has thrown considerable weight upon the evidence of the introductory pages and upon cartographic testimony.

It has been shown that various states exist for the introductory pages and for three of the maps ('Anglia ...', Norfolk, and Northamptonshire). In the case of the engraved frontispiece it has been conjectured that State II appeared early, possibly even in 1579. The dating of the various settings of the index is similarly difficult, although a priority with respect to one another may be fairly definitely established. And it can be reliably determined that the table of coats-of-arms was executed after September, 1589. The addition of the marginal graticule to the map of 'Anglia ...' might be associated with Saxton's large map of England and Wales (1583) which is so graduated (see later section); and the addition of names to the two county maps might also relate to this date, the map of 1583 possessing the same names. It is reasonable to suggest that by tabulating the states of the introductory pages and the three maps in various copies of the Atlas, and by noting whether or not the coats-of-arms occur, then it ought to be possible at least to place certain issues in chronological sequence with respect to one another. Such tabulated information is detailed in Appendix 10, but it is to be noticed that the results are as confusing as one might have hoped they would be helpful. Copy No.7, for example, possesses the frontispiece in State II and each of the three maps in State I; yet Copy No.45 has the frontispiece in State I and two of the maps in State II. The most that can be presumed is that certain copies, e.g. Nos.5 and 36, were among the earliest, while copies which had assumed their definitive form, e.g. Nos.4 and 6, were among the latest, *c.*1590. The confusion inherent in this method of dating the various copies results from the possibility that some of the copies might have been assembled from sheets acquired separately and in various states, and the practice of sophistication, i.e. the insertion or substitution of various sheets from other copies for one reason or another.[8] However Copy No.45, on vellum, must have been printed as an entity; it is impossible to date this particular issue, but it nevertheless follows that the two county maps had already been altered while the frontispiece and the general map were still in their original state. No satisfactory method, therefore, exists to establish a chronological sequence for the appearance of the various issues of Saxton's Atlas. Neither watermarks, quality of impression, nor cartographic evidence point clearly in any one particular direction, although the general evolution of the atlas can nevertheless be traced up to c.1590.

(d) The Contents and Format of Saxton's Atlas

Saxton's Atlas consists of introductory leaves and maps. There appears to have been no set sequence for the three introductory pages, though the frontispiece, where it appears, usually occurs first. The index and the coats-of-arms/table of cities are variously located in different copies, although usually at the beginning of the Atlas. Saxton's maps occur in a sequence corresponding to that in the setting of the index appropriate to any particular copy. Exceptions to this rule are so rare that they almost certainly represent errors in assembly and binding. All map titles are in Latin.

Additional but not necessarily contemporaneous maps are interpolated, on occasion, into certain copies of the Atlas (see Appendix 10). For example, the British Library copy (Map Room; c.7c.1) has eleven maps added at the end, showing the engagements of the Spanish Armada and English Fleet (*Expeditionis Hispanorvm In | Angliam vera descriptio | Anno D: MD LXXXVIII*. Engraved by Ryther to accompany P. Ubaldini's *Discourse*). Another copy in the British Library (Appendix 10, Copy No.6) contains several maps relating to Scotland and Ireland, together with the Quartermaster's map by Hollar, 1752.

Saxton's maps number thirty-five; one is the general map of England and Wales ('Anglia...'), and the remainder thirty-four county maps depicting fifty-two counties. Twenty-four are maps showing individual counties, while ten maps show county groups covering in total twenty-eight counties. The maximum number of counties included on any one sheet occurs on the map of Northamptonshire/Bedfordshire/Cambridgeshire/Huntingdonshire/Rutland, though four counties are shown on the maps of Kent/Sussex/Surrey/Middlesex, and Radnor/Breconshire/Cardiganshire/Carmarthenshire. All other maps of county groups (excepting that of Oxfordshire/Buckinghamshire/Berkshire) show two counties each (Table 3). Constituent maps vary slightly with regard to dimensions, the average being 397×510 millimetres (measurements are given between the outer limits of the printed surface). With the exception of the large map of Yorkshire, the height of the maps varies from 346 millimetres (for the maps of Norfolk and Suffolk) to 428 millimetres (for the map of Wiltshire). The length of the maps similarly ranges between 445 millimetres (Hampshire) and 564 millimetres (Dorset). The map of Yorkshire is exceptional, measuring 545×743 millimetres, and as a result is usually folded to permit its inclusion in various copies of the Atlas.

All the maps have a conventional orientation, north being uppermost, though occasionally there may be a slight adjustment to allow a particular area to be conveniently fitted onto the sheet, e.g. the map of 'Anglia...' (Plate 6) has its north point displaced somewhat towards the north-east. The calculation of such minor adjustments, however, is extremely difficult, as only one map (Devon) possesses a compass-rose. In this case north has been slightly displaced towards the north-west. All the maps, however, possess within or close to their borders Latin words defining the four cardinal directions, i.e. 'SEPTENTRIO', 'MERIDIES', 'OCCIDENS', and 'ORIENS'. With the exception of only three maps, the words invariably occur enclosed within the border of the maps. They are always (with the exception of the map of Hertfordshire) in upper-case lettering, and occur in cartouches of slightly variant but always simple design (though on the maps of Somerset and Warwickshire these cartouches are absent, and on the maps of Hertfordshire and Northumberland scrolls are used instead). The map of Cornwall, however, shows each word printed on the face of the map itself, without any containing cartouche or scroll. Indeed, there is the suggestion that they may have been added rather hurriedly at some stage subsequent to the completion of the plate itself; the lettering of 'SEPTENTRIO' and 'OCCIDENS', for example, is carelessly applied, and the former is not centrally positioned along the upper margin of the map.

Other than the cardinal points, no information of any direct cartographic import is contained in the map borders. These vary in width from 5 millimetres on the map of Cornwall to, most frequently, about 8–10 millimetres. Almost all the borders consist of an extremely narrow inner and outer blank margin, and an intervening space filled with decoration of various types; engravings of leaves are commonly used in this capacity, though different designs also occur. The only exceptions to the above generalisation are the borders of the map of Hertfordshire (in which no decoration appears), those of the map of Warwickshire (which lack any narrow blank margins), and those of the map of 'Anglia . . .' which, in the map's second state, incorporate a graduation of latitude and longitude (Plate 6).

Titles of all the maps are in Latin (Table 3). They occur in conveniently located cartouches embellished according to the style of the engraver who executed them. Thus, for example, the title of the map of Kent is almost inconspicuously inserted in the top right-hand corner of the map in a small, plain rectangle, surrounded by the minimum of decoration. Similarly, the title-cartouches of the maps of Denbighshire and Shropshire consist of plain double-bordered rectangles outside which there appears uncomplicated decoration. Some maps, e.g. Cheshire, reveal ornate but not superfluous ornamentation. By contrast, the title-cartouches on the maps of Durham, Northamptonshire, and Suffolk are highly embellished, with birds, bees, beetles, flowers, and human figures, amongst other features, providing rich adornment.

Each map possesses the Royal Arms of Queen Elizabeth, with supporters on all but four (Kent, Northamptonshire, Somerset and Suffolk). More often than not they surmount the title-cartouche, although on twelve maps (Anglesey, Derbyshire, Devon, Durham, Kent, Lancashire, Lincolnshire, Norfolk, Oxfordshire, Staffordshire, Westmorland and Worcestershire) the arms occur separately from the title. Reasons for locating such features in varying map locations are essentially questions of cartographic design and balance, and do not require analysis. Ten of the maps have the Royal Arms surmounted by the cypher '.ER.' ('Anglia . . .', Cornwall, Durham, Essex, Glamorgan, Hampshire, Oxfordshire, Wiltshire, Gloucestershire and Yorkshire). In the case of two others (Devon and Lincolnshire) this is incorporated into the decorative framework of the Royal Arms; and on the map of Westmorland the Royal Arms are surmounted by the words 'Elizabeth Regina'.

The arms and motto of Thomas Seckford of Woodbridge in Suffolk, Saxton's patron, appear together on all the maps, placed in any position which does not involve the omission of significant cartographic detail. As previously noted (Chapter 2), sometime during 1576 Seckford's motto was changed from 'Pestis patriae pigrities' to 'Industria naturam ornat'. Twenty-four of the maps carry, as they did in their original proofs, the later motto (though those of Durham and Warwickshire also possess the former). The earlier motto appears on thirteen of the maps (including those of Durham and Warwickshire), having been added to the maps of Kent, Oxfordshire, Norfolk, Suffolk and Somerset subsequent to the appearance of the Burghley proofs.

Names of the engravers of various maps are not always given, being omitted from thirteen maps (Table 1). In the case of Suffolk, however, it is known that the engraver was Lenaert Terwoort; his name, which had appeared on the Burghley proof, had since been replaced by Seckford's earlier motto, there being little alternative space for its inclusion. Where the engravers' names do occur they are in almost all instances incorporated into cartouches or scrolls associated with various other features on the map, e.g. line-scales, title-cartouches, or open dividers.

The acknowledgement 'Christophorus Saxton descripsit' (the Christian name of the cartographer being variously spelt) occurs on every map, in twenty-four instances having been added subsequent to the appearance of the Burghley proofs (Table 1), in positions convenient for its late insertion. Style suggests that such late addition onto various maps was probably effected by a single engraver, and almost certainly by a person other than the engraver of the rest of the map.

Open dividers occur on all maps in the Atlas, and in each instance they surmount the line-scale. The majority are of extremely simple design, consisting of two arms and a circular hinge at their apex. Four maps however, each engraved by Ryther ('Anglia...', Gloucestershire, Westmorland and Yorkshire), possess dividers of more elaborate design, the two arms crossing one another before linking in a decorative hinge. In almost all cases the dividers stand centrally or nearly so upon the line-scale.

Legends do not occur on Saxton's maps, although explanatory panels appear on five. That on the map of 'Anglia...' is an 'Index' to the counties appearing on the map. Norfolk possesses a similar index to the hundreds in the county. And the maps of Hampshire, Oxfordshire and Kent have panels containing geographical information of a general nature regarding the counties covered on each of the three maps.

Line-scales of varying style and length appear on all the constituent maps of Saxton's Atlas. They are usually inserted in some convenient but otherwise unimportant space on the page, for example areas of sea, or of counties outside those which form the main subject of the map itself. The line-scale on the map of Kent, for example, occurs in the English Channel, while on the map of Herefordshire it occupies a position in the bottom left-hand corner (in Breconshire). The style of these scales is comparatively simple, consisting often of nothing more than mile graduations along a horizontal bar and, not infrequently, subdivisions of miles. It has been demonstrated by various researchers that the mile used by Saxton was roughly equivalent to 1·3 statute miles.[9] These line-scales are always surmounted by a pair of open dividers (see above) and by the words 'Scala Miliarium' (although on the map of Hertfordshire the words appear below the line-scale). Frequently the line-scale and its accompanying dividers provide convenient positions for an acknowledgement of Saxton as author of the survey and for the name of the engraver. Map scales vary between 1/313,783 (Kent, Lancashire, Lincolnshire and Northumberland) to 1/146,432 (Table 4).[10] Saxton clearly used widely divergent scales on his maps, the extent of the area to be mapped being the principal criterion. Not surprisingly, therefore, most of Saxton's county groupings occur as small-scale maps.

(e) The County Maps

The main geographical features shown on the county maps comprise relief, including water; vegetation; and cultural features such as settlements, large houses, and other notable buildings. The boundaries of the hundreds are occasionally shown (on five maps only).

Hills and mountains are pictorially defined. They vary in size not so much according to the scale of the map as according to the general elevation of the summits and landscapes represented. Thus the downs of Sussex are shown less dramatically than the hills of Northumberland, though both maps are at an identical scale. The aim was to convey an impression of topography rather than to provide precise information on the location and altitude of individual summits. Where particularisation was felt necessary, hills were named, e.g. 'Plymllymon hill' and 'Caddorydric hill' on the map of Montgomeryshire.

The majority of hill-symbols are shaded on their south-eastern slopes, though exceptions occur. The significance of Saxton's hill-symbols has been the subject of speculation in the past. Certain hills appear to be considerably and unjustifiably more elevated than others and some are named; field-observations in northern England suggested to Manley that such hills might have been climbed by Saxton or an assistant and used as surveying points.[11]

Expanses of sea, where they occur, are shown as closely stippled areas. These frequently provide space for the inclusion of incidental features on the maps, e.g. the arms of Queen Elizabeth and of Thomas Seckford on the map of Lancashire. Elizabethan ships, fishes, and sea-monsters frequently occupy areas of sea which might otherwise appear empty and bare. Lakes, similarly, as on the map of Westmorland, are often stippled.

Rivers are usually shown as a series of closely drawn sinuous lines following the course of any particular stream. Saxton's portrayal of general drainage systems was fairly good, but his attention to detail with regard to the location of small settlements and similar features relative to minor streams was often at fault. Considering the speed at which the survey was effected, however, such a situation is hardly surprising.

Woodlands are shown conventionally by small pictorial tree-symbols which have often been painted green subsequently. The size of these symbols varies proportionately to the scale of the map. They are sometimes located in clusters and named, e.g. 'The New forest' (Hampshire), though on some maps the tree groupings are less conspicuous despite the occurrence of names, e.g. 'The forest of Sherewood' on the map of Lincolnshire. Parklands are more precisely identified, being enclosed by ring-fences and sometimes, though by no means always, being named. Such areas of parkland were perhaps best known because they surrounded a large country house, possibly the seat of some county notable, and these occasionally appear within the park confines. Where woodland occupies part of the parkland area, tree-symbols are appropriately inserted to convey the correct impression.

A variety of representation is employed to show buildings and settlements. The small symbol of a building surmounted by a spire frequently defines a village, while more important towns are often represented by groups of buildings, sometimes with a church-like edifice in their midst. As the scale of the maps increases, so too does the degree of detail with which settlements are defined. It would seem that the representation of cultural features of this kind was effected according to a carefully thought out system, although the difficulties of engraving small symbols inevitably mean that on occasion slight variations occur. Almost all settlements are named on the maps. Usually the names of the more important centres are printed in capitals, while the remainder are in upper and lower case. Only rarely does one discover a symbol without an accompanying name, or vice-versa, and such instances would appear to be cases of careless omission by the engraver.

Most of the detail shown on the maps refers to territory within the county boundaries to which any particular map directly refers. The names of adjacent counties are almost always given in appropriate positions on the maps, however, and a selection of hills, parks, woods and settlements in such counties also appears. Otherwise the adjacent counties often provide space for the insertion of features such as the title-cartouche or coats-of-arms, there being in these parts of the map considerable space for their inclusion.

The political framework to which each of Saxton's maps is appended is the county structure of England and Wales. Other than the limits of the counties themselves, the boundaries of hundreds are shown on only five maps (see Chapter 2a). In each instance these are represented by dotted or pecked lines. On the map of Norfolk the hundreds are named in a small panel in the top right-hand corner of the map, while capital letters occur alongside the list and also on the face of the map, providing a convenient method of reference. On the four other maps the names of hundreds are printed in place on the face of the map itself, always in capital letters. This practice of showing hundreds was extended by later publishers of the Atlas (see Chapter 5) until eventually every map possessed them, e.g. William Web, in 1645, added their names and divisions to the map of Lincolnshire and their divisions (without names) to the map of Yorkshire.

(f) The map of England and Wales ('Anglia . . .')
It would seem (Chapter 4) that the map of 'Anglia . . .' (Plate 6) probably derived from another map of similar character, such as Mercator's of 1564. There is, at least, good reason for supposing that it was not merely a generalised reduction of detail extracted from the county maps. Nevertheless, the geographical information which it shows, while more generalised and less particularised, is essentially the same as that given on the larger-scale county sheets. The map, as already mentioned, occurs in two states, the latter showing a marginal graduation of latitude and longitude (Chapter 3b). The Latin title occurs in a heavily embellished cartouche in the top right-hand corner, surmounted by the Royal Arms with supporters. Below this are Seckford's arms and later motto, while the line-scale with open dividers appears to the north-west of the Cornish peninsula. The top left-hand corner of the map is taken by an 'Index Omnivm Comitatvvm'; this comprises a list of the counties with reference numbers to their right, and facilitates the identification of the counties which are unnamed on the face of the map. Beneath this, in a small decorated cartouche, there is a Latin note relating to the character of the towns of England and Wales. Sea-areas are occupied by numerous ships, sea-monsters and fishes, and are stippled as they are on the county maps, while areas of sand-flats are indicated by peck-marks offshore. Four semi-circular compass-roses are attached midway along each of the edges of the map, a feature peculiar to this sheet, but the cardinal points are still printed, in Latin, in the map's borders.

Hills and mountains are again pictorially defined, though symbols are considerably fewer than on the larger-scale county maps. Shading on the south-eastern slopes remains clearly visible even at this scale. Rivers are shown in reasonable detail, though some generalisation has occurred and only rarely are streams named. Woodlands are represented as on the county maps, though in less detail, but parklands are omitted.

Considerably fewer settlements are shown than appear on the county maps. Important towns are always marked, however, and are usually named in capital letters, whilst a selection of less important centres is also included. Hundreds are excluded from the map, and detached portions of counties, which were detailed on the county sheets, are no longer shown. County boundaries are clearly marked as dotted lines, although they are less conspicuous where rivers form the borders.

(g) Colour on Saxton's maps
Most copies of Saxton's Atlas and, indeed, of its later editions (see Chapter 5), reveal the addition of colour to the maps. This was the work of individual owners and does not follow any universal pattern. Colour was applied largely for visual attraction rather than functional improvement, though a minority of copies (e.g. Appendix 10, Copy No.6) were not amended in this way. The current popularity of old maps probably owes much to this subsequent addition of colour, but it is unfortunate that the fine line-work and lettering which epitomises so much of the engraving of this period has been so often and so heavily obscured by the practice. Apart from a certain superficial attractiveness and occasional practical advantage there is little to be gained from painting on copperplate engraving.

(h) *Saxton's wall-map of 1583*[12]

Not so well documented as Saxton's Atlas is the large wall-map of England and Wales which he issued in 1583, entitled *Britannia Insularem in Oceano Maxima*. Being more exposed to the ravages of time than maps in atlases, few maps of this kind have survived and Saxton's large map of England and Wales is no exception. The British Library possesses a copy in the form of a folio volume (Maps c.7d.7) containing the twenty sheets of which the wall-map was composed; but as the watermark on most of these sheets was one current for a limited period only in the mid-seventeenth century, it seems that this was a copy reprinted without change (a 're-strike') about 1640.

In 1939, however, the Birmingham Public Libraries obtained a copy which appears to be in the original state (catalogue No.493213), differing from the British Library's copy in possessing an ornamental border exhibiting coats-of-arms of the English nobility. The watermark of the constituent sheets is the bunch of grapes found in early issues of the Atlas, implying an early if not original version. The Birmingham map, unlike that in London, has been assembled into a single picture of the entire country and mounted on canvas, and therefore occurs in the form for which it was initially designed.

It may be supposed that much of the geographical information shown derived from Saxton's national survey, but at the smaller scale (approximately 200 millimetres to one mile) is more selective and generalised than similar detail presented in the Atlas. In other respects the similarities between this and the Atlas maps are numerous. The engraving is intricate and embellished, Elizabethan ships and other decorations occurring in the surrounding seas. No reference is made to the engraver, however, and authorship can only be supposed on the grounds of style. In particular, the open dividers recall the work of Augustine Ryther, who engraved several of the county maps in the Atlas, together with the map of England and Wales ('Anglia . . .').

(i) *The engravers of Saxton's maps*

Christopher Saxton's Atlas was issued during that period when Dutch and Flemish cartographers dominated the world of map-making. The celebrity of Ortelius and Mercator, amongst others, had not merely asserted the supremacy of the north European over the by now relatively declining Mediterranean centres of cartography, but had established the fashion in cartographic design and engraving which was to influence map-making for generations to come. Thus it is unsurprising to witness the contemporary emergence of the first English national atlas, for the Dutch and Flemish were pre-eminently the creators of the atlas as we now comprehend it.

Saxton, as was to be expected, employed some engravers who were themselves of Dutch or Flemish origin; Remigius Hogenberg, Lenaert Terwoort, and Cornelis de Hooghe engraved between them fifteen maps in the atlas. Remigius Hogenberg (nine maps), a maker of scientific instruments and native of Mechelen, was brother to the equally famous Frans. He cannot have arrived in England much before 1572, when he engraved a portrait of Archbishop Parker, because he issued a large *View of Munster* in 1570. He remained in England until 1587, when he engraved John Hooker's plan of Exeter. Less is known of Terwoort (five maps), a native of Antwerp. His only work which has come to light, other than that for Saxton, is an illustration dated 1591 in Thomas Hood's *Making and Use of the Sector*, 1598. De Hooghe is known to have been a pupil of Philip Galle at Antwerp. His sole contribution to Saxton's Atlas and his only known work in England was the map of Norfolk (1574). He claimed to be a natural son of the Emperor Charles V, and was executed in 1583 for implication in a plot against the life of William the Silent. But Saxton also employed Englishmen whose training and expertise reflected the style of their

Continental masters. Augustine Ryther, one of the earliest English exponents of copper engraving, has been described by Ralph Thoresby[13] as 'probably of Leeds', and he was possibly an offshoot of the ennobled Yorkshire family of Ryther. Nothing is known of him prior to his work for Saxton, and it is reasonable to suggest that an aquaintanceship with the latter might have promoted his career. He engraved five of Saxton's maps and possibly the frontispiece to the Atlas. Hind,[14] however, believed that the latter was more likely to have been the work of Hogenberg, who had already produced some fine portraits the arrangement of which reflects that in Saxton's Atlas; Ryther, by contrast, showed neither the inclination nor ability to engrave human forms. Certain stylistic similarities, particularly the form of the open dividers above the line-scale, suggest that Ryther may also have been responsible for engraving Saxton's wall-map of 1583, but no positive evidence exists to support the hypothesis. He engraved several works, including the Armada Plates (see Chapter 3a), but for a craftsman of such attainment came to an ignominious end, nothing more being heard of him after the winter of 1594-5 when he was in the Fleet Prison in London. Francis Scatter, known only through his work for Saxton, engraved two of the Atlas maps, and Nicholas Reynolds that of Hertfordshire.

Between them these engravers account for twenty-two of the maps found in the Atlas, the remaining thirteen being unsigned and therefore unattributable (Table 5). Among the latter, however, the proof map of Suffolk in Lord Burghley's collection contains the name of Lenaert Terwoort, subsequently to be erased and replaced (see Table 1, footnote 5). Hind[15] has claimed that Burghley's proof copy of Radnorshire indicates that the map can be attributed to Francis Scatter, but the sheet reveals neither his name nor any justification for the hypothesis. Similarly Skelton[16] has said that the words 'Johannes Rutlinger (sculpsit?)' are imperfectly erased from the plate of Oxfordshire, having originally appeared at the bottom of the title-cartouche, but there appears to be no evidence for this. It is occasionally possible to assign an unsigned map to a particular engraver on the grounds of style; the map of Northumberland, for example, possibly pertains to Ryther. Lynam[17] has surmised that Terwoort might have engraved the map of Northamptonshire, if one is to judge by the characteristic birds, bees, and butterflies in its cartouche. It has also been suggested that Saxton himself engraved certain sheets, though the evidence for this is weak and hardly conclusive; Walpole[18] stated that 'Saxton published a compleat set of maps of the Counties of England and Wales, many of which he engraved himself'. If true this would account for some of the twelve unattributed maps in the Atlas. Hind,[19] however, fails 'to find any special quality or peculiarity to distinguish between the various engravers in Saxton's circle', and therefore prefers not to attempt to assign unsigned maps to any particular individual.

TABLE 3

Saxton's Atlas. Titles of maps

Sequence as in the definitive form of the Atlas, c.1590, corresponding to that in Setting D of the Index.

1. ANGLIA / hominu numero, rerum' fere / omniu copiis abundans, sub mi= / tissimo Elizabethae, ferenissimae / et doctissimae Reginae, imperio, / placidissima pace annos iam / viginti florentissima. / Anº Dm / 1579.

2. CANTII, Southsexiae, Sur / riae, et Middelsexiae comitat', / Una cum suis vndique / confinibus, Oppidis, pagis, / Villis, et fluminibus, in / cisdem, vera descriptio. / Anº Dm 1575.

3. SOVT / HAMTONIAE / Comitatus (preter Insulas / Vectis, Jersey, et Garnsey, / quae sunt partes eiusdem / comitatus) cum suis vndiq, / confinibus; Oppidis; pagis; / Villis; et fluminibus; / Vera descriptio.

4. DORCESTRIAE / Comitatus Vicinarumque / Regionum noua veraq Descriptio / Anno Dni 1575.

5. WIL / TONIAE / Comitatus (herbida / Planitie nobilis) / hic ob oculos pro; / ponitur. Anno / Dñi 1576.

6. SOMERSETENSEM / Comitat' (agri fertilitate / Celebrem) hec ob oculos / ponit Tabula. / Anno 1575.

7. DEVONIAE COMITAT', RERVMQVAE / omnium in eodem memorabilium re: / cens, vera pticularis descriptio. / Anno Dn 1575.

8. PROMONTORIVM HOC / IN MARE PROIECTVM / CORNVBIA DICITVR.

9. ESSEXIAE / COMITAT' NOVA / vera ac absoluta descriptio / Ano Dni / 1576.

10. HARTFORDIAE / COMITATVS noua, uc = / ra, ac particularis descrip: / tio. Anno Dni. 1577.

11. Oxonii buckinghamiae et berceriae / Comitatuum, Vna cum suis Vndiq, / Confinibus, Oppidis, pagis, villis, / et fluminibus in eisdem vera / descriptio. Ano Dñi 1574.

12. GLOCESTRIAE / Siue Claudiocestriae Comitat' / (Claudij Caesaris nomine ad huc celebrat') / Verus Tipus atq effigies. / Ano Dñi 1577.

13. SVFFOLCIAE Comitatus continens in se' / Oppida mercatoria :25: Pagos et Villas: 464 / Vna cum singulis Hundredis et fluminibus / In eodem Vera descriptio: / Anno Domini 1575.

14. NOFOLCIAE / comitatus continens in se'. Oppida / mercatoria 26, Pagos et Villas 625, / Una cum singulis Hundredis, et flu / minibus in eodem, Vera descriptio.

15. NORTHAMTON / Bedfordia' Cantabrigiae' / Huntingdoniae et Rutlandiae' / Comitatuum Vicinarumq / regionum partium adiacent' / noua veraq descriptio / Ao D. 1576.

16. WARWIC / LECESTRIAEQ / Comitat': Ciuitat': / Oppidorū. Villarū. / fluminū. Ceterarumq/ rerum omnium in / eisdem memorabi = / lium. noua. Veraq / descriptio.

17. STAFFORDIAE / Comitatu' pfecte et / absolute elaboratu haec / tibi tabula exhibet / Anno Dni 1577.

18. WIGORNIENSIS / Comitatus Sabrinae / Fluminis Amaenitate / insignis descriptio / Ano Dñi 1577.

19. SALOPIAE COMITATVS / summa cum fide, cura et dili: / gentia descriptionem haec ti: / bi tabula refert. Ao Dñi 1577.

20. FRVGIFERI / AC AMENI / HEREFORDIAE / COMITATVS / DELINIATIO / Anno Dñi 1577.

21. LINCOLNIAE NOTINGHAMIAEQ / Comitatuu noua vera et / accurata descriptio. Anno / Domini 1576.

22. VNIVERSI / Derbiensis Comitatus / graphica descriptio 1577.

23. CESTRIAE / Comitatus (Romanis / Legionibus / et Colonys / olim insignis) vera et / absoluta effigies.

24. EBORACENSIS / Comitatus (euius Incolae olim / Brigantes appellabantur) Lon: / gitudine Latitudine hominu / numero reliquis illustrior. / Ano Dm 1577.

25. LANCASTRIAE / Comitatus palatin' vera / et absoluta descriptio / Anno Dni 1577.

26. DVNELMENSIS / Episcopatus, Qui comitatus / est palatinus vera et / accurata descriptio / Ano Dni 1576.

27. WESTMORLANDIAE / et Cumberlandiae Comit' / noua vera et Elaborata / descriptio. Ano Dni 1576.

28. NORTHVMBRIAE / COMITATUS / (Scotiae contiguae) / Noua Vera / descriptio.

29. MONVMETHENSIS / Comitatus Regis / Henrici quinti / natalitiis celeberrimus / Anº Dñi 1577.

30. GLAMORGĀ / Comitatus, australis / Cambriae pars descriptio / Anº Dñi 1578.

31. RADNOR BREKNOK / Cardigan et Caermarden / quatuor australis Cambriae / comitatuum (B. Dehenbart. A. / Southwales) descriptio. Anº / Dñi 1578.

32. PENBROK / comitat' qui inter meridionales / cambriae Ptes hodie censetur olim / demetia. L Dyfet' ptes B hoc est / occidentalis wallia descriptio / Anº Dñi 1578.

33. MONTGOMERI ac / Merionidh, duorum borialis / cambriae comitatuum. B. / Gwinedhia. A. Northwales / nuncupat descriptio Anº Dñi 1578.

34. DENBIGH / AC FLINT / duorum olim / cambriae, / modo Wal: / liae, comi: / tatuum. / descriptio / Aº. Dño. 1577.

35. MONE INSVLAE / modo Anglesey, et Caernaruan, / duorumborialis cambriae comitatuu' / olim uene docia. L. Gwynedhia / B. Northwales. A. descriptio, / Anº Dñi 1578.

TABLE 4

The scales of Saxton's county maps

Map	Scale
Kent, etc. Lancashire Lincolnshire, etc. Northumberland	1/313,783
Cornwall	1/304,128
Devonshire	1/299,520
Yorkshire	1/282,404
Radnorshire, etc.	1/274,560
Oxfordshire, etc. Westmorland, etc.	1/263,577
Warwickshire, etc.	1/255,075
Anglesey, etc. Hampshire Northamptonshire, etc. Staffordshire	1/253,440
Gloucestershire	1/247,104
Montgomeryshire, etc. Norfolk Suffolk	1/235,337
Somerset	1/219,648
Derbyshire	1/205,920

Map	Scale
Shropshire	1/202,752
Worcestershire Pembrokeshire	1/193,807
Essex	1/188,269
Dorset	1/183,040
Cheshire	1/181,777
Denbighshire, etc. Glamorganshire	1/178,024
Durham	1/175,718
Wiltshire	1/173,406
Herefordshire	1/171,898
Hertfordshire	1/156,894
Monmouthshire	1/146,432

TABLE 5

The engravers of Saxton's maps in the Atlas of England and Wales, 1579

1. *Remigius Hogenberg*
   Denbighshire, etc.; Devonshire; Herefordshire; Kent, etc.; Lancashire; Lincolnshire, etc.; Montgomeryshire, etc.; Shropshire; Wiltshire.

2. *Lenaert Terwoort*
   Cornwall; Hampshire; Somerset; Suffolk[1]; Warwickshire, etc.

3. *Cornelis de Hooghe*
   Norfolk.

4. *Augustine Ryther*
   Anglia; Durham; Gloucestershire; Westmorland, etc.; Yorkshire.

5. *Francis Scatter*
   Cheshire; Staffordshire.

6. *Nicholas Reynolds*
   Hertfordshire.

1. Indicated only on Lord Burghley's proof. Replaced in Saxton's Atlas by the motto of Thomas Seckford in its earlier state, i.e. 'Pestis patriae pigrities'.

Chapter 4

Saxton's sources of information and contemporary surveying practices

Saxton's maps were probably compiled both from inquiry into existing cartographic and documentary sources, and from information obtained in the field. The county maps, almost certainly, derive to a considerable extent from field observations, though existing source information must have been consulted up to a point. The map of 'Anglia ...', however, being at small scale, was more the result of generalisation and selection from other maps.

Most of the extant maps relating to England and Wales which survive from times prior to Saxton's national survey cover the country either as a whole or in large regional sub-divisions, and are considerably too generalised to have served as important source material for Saxton's basic work. Laurence Nowell produced a small-scale map of the British Isles in c.1563 (*A general description of England & Irelãd with ye costes adioyning*), but it lacks the detail of Saxton's county maps or his map of 'Anglia ...'. Similarly, Mercator's map of the British Isles (*Angliae Scotiae et Hiberniae noua descriptio*, 1564) is at far too small a scale to have been of much use in the compilation of Saxton's maps; it shows little more than county names and some important towns. Despite this, both Saxton's map of Yorkshire and Mercator's map reveal certain similar name omissions, and the possibility of some relationship cannot be ruled out.[1] Indeed, it seems very probable that Mercator's map was based on information supplied by John Rudd, who was most likely Saxton's teacher and mentor, compounding the likelihood of this linkage.[2] Lynam[3] has suggested two collections of regional maps, now vanished, which Saxton might have consulted. One of these belonged to William Lambarde, the Kentish antiquary, who inherited many of Nowell's maps and who acquired or drew other local maps while compiling his great *Topographicall Dictionarie* of England between c.1568 and 1577. Again, Raphael Holinshed tells us in his *Chronicles* that Reynold Wolfe, the initiator of that work, 'spent a great part of his time' on maps. Wolfe, the Royal printer, had already announced that maps of the English provinces would illustrate the *Universall Cosmographie* which, with Holinshed's assistance, he was preparing but which he never completed. Wolfe, however, died c.1573, and the maps upon which he had been working may have been placed at Saxton's disposal, probably at Seckford's request. William Harrison, in his *Description of Britaine* (printed in the first edition of Holinshed's *Chronicles*), said of the counties: 'And these I had of a friende of myne, by whose traveyle and hys maisters excessive charges I doubt not, but my country men eare long shall see all Englande set foorth in severall shyres after the manner that Ortelius hath dealt wyth other countries of the mayne, to the great benefite of our nation and everlasting fame of the aforesayde parties'. Perhaps Saxton's Atlas, derived partly from details assembled earlier by Wolfe, would now illustrate and to some extent complement Holinshed's *Chronicles*. Saxton might also have consulted Leland's *Itinerary*, either directly or indirectly through Reynold Wolfe. Wolfe had acquired many of the manuscripts of John Leland after the latter's breakdown in health; and although Lynam does not believe that Leland ever drew any maps except for a few rudimentary sketches, his *Itinerary* may have inspired Wolfe to collect maps for inclusion in the *Universall Cosmographie*. Manley,[4] however, on the basis of map content, considers it unlikely that Saxton referred to Leland. It has been suggested by Bedford[5] that the five so-called Weston tapestry maps, dated as later than 1570 (and definitely at 1588 in one instance), are in many cases very similar to the maps of Saxton and possibly derive from a common source, but the latter remains unidentified. While one cannot be specific, it seems indisputable that Saxton should have used some contemporary cartographic and documentary sources to assist in the preparation of the county maps. Such material was available at the time, despite the fact that it was considerably inaccurate and inadequate. These limitations, however, could presumably be made good through personal investigation, and it was probably field observation of this type with which Saxton was primarily concerned during his national survey rather than dimensuration of a totally original and exhaustive nature.

Compilation of the general map of England and Wales ('Anglia . . .'), unlike that of the county maps, was almost certainly achieved by generalisation and selection from existing maps. Saxton possibly referred to his county maps in certain instances, but there are features on the map of 'Anglia . . .' which suggest other sources too. The map shows, for example, not only adjacent parts of Ireland, Scotland, and France, but it includes a selection of towns in each unobtainable from the county maps. The Isle of Man is included, with places such as 'Dowglas' and 'The calfe of man' clearly identified. Part of the Isle of Wight is included on the map of Hampshire, but its northern coastline is drawn in a manner considerably less accurate than that on the map of 'Anglia . . .' (where the island is shown in its entirety); and the county map shows no settlements on the island, unlike the national one. Placenames, too, reveal disparities. In the counties of Hampshire and Sussex, for example, all those places marked on the map of 'Anglia . . .' appear on the relevant county maps (excluding Newport and Ryde, shown in the Isle of Wight on the map of 'Anglia . . .'). But the spellings vary considerably between national map and county sheets. Such a feature would have been unlikely had the county maps been rigorously consulted during the compilation of the former. Similarly, whilst the Isle of Wight is referred to as 'WIGHT INSVL' on the county map of Hampshire, it becomes 'VECTUS INSVL' on the map of 'Anglia . . .'. More likely than compilation directly from the county maps would be reference by Saxton to existing small-scale maps of the British Isles or parts thereof. North[6] has closely compared Wales on the map of 'Anglia . . .' with various sources available at the time, but has come to the conclusion that Saxton's work was in the main original. As far as the coasts and rivers are concerned, Saxton's map was a great advance on those of Mercator (*Angliae Scotiae et Hiberniae noua descriptio*, 1564) and Lhuyd (*Angliae Regni Florentissimi Nova Descriptio . . .* 1573 and *Cambriae Typus . . .*), particularly in South Wales where, for the first time, Gower and St Bride's Bay appear. Indeed, with regard to natural features, Saxton's map does not appear to have been influenced by any other except in two respects. One of these relates to the change in the trend of the Glamorganshire coast at Lavernock Point, suggested on only one other map, i.e. Mercator's. Second, the islands (actually non-existent) which are drawn off the southeastern coast of Ireland on Mercator's map appear in almost identical shape, number, and position on Saxton's. The only earlier map in which any such features are suggested is Lhuyd's map of Wales, but here they differ both in number and position. Generally, Saxton's names do not correspond at all closely to those on earlier maps, although there are some similarities between his map and that of Mercator. There is obviously a reasonable possibility that Saxton may have referred to Mercator's map, but from the evidence so far assembled it is impossible to be positive, and much of the information contained on his map of 'Anglia . . .' most likely derived from first-hand experience. The wall-map of

1583 seems to have its origins in Saxton's survey for the Atlas, but such a direct link is hardly surprising.

Saxton's training as a surveyor equipped him well for an undertaking of such magnitude. Fordham,[7] amongst others, has assembled numerous documentary references in which Saxton is specifically mentioned in this capacity. The revolution in surveying which occurred between 1550 and 1650 was associated with three contemporary developments:[8] (a) the increasing interest in navigation, and its accompanying application of astronomical theory; (b) progress in the military sciences coupled with the application of geometry to problems of sighting, i.e. in gunnery, and the employment of gunsmiths in the production of experimental instruments; (c) the agrarian changes of the period. The replacement of the open fields meant that land boundaries had to be accurately defined. A large number of treatises began to appear on surveying, and their popularity is demonstrated by the fact that they were so often reprinted. The lack of any general training in mathematics exerted considerable influence on the character of these publications and on the construction of early surveying instruments.[9] This factor was significant in the design, for example, of the plane-table,[10] an instrument which would perform more or less mechanically the numerous arithmetical calculations involved in land measuring, and the early development of which is represented by the holometer of Abel Foullon.[11] This lack of mathematical proficiency resulted in instruments with scales and indices to assist in the determination of inaccessible heights and distances, and such scales remained in almost constant use until trigonometry became more widely known and applied in the early eighteenth century. Geometry was even less widely understood than arithmetic, though its relevance was appreciated by Edward Worsop,[12] who documented four errors frequently committed by surveyors: (a) multiplication of length by width to give area, without any consideration as to whether the angles between the four sides were right-angles; (b) estimation of angles by eye, without the assistance of an instrument; (c) estimation of the area of irregular figures through adding the lengths of their sides, dividing by four, and squaring the result; and (d) estimation of the area of a circular figure through dividing its circumference by four, and squaring the result. This deficiency in mathematical comprehension explains why so many introductory pages of contemporary surveying texts were devoted to explanatory passages. The first five chapters of Digges' *Tectonicon*,[13] for example, deal with principles of arithmetic and geometry. The remainder relate these principles to surveying practice, and discuss the use and application of various instruments. These are the geometrical square on the side of a ruler, the carpenter's square, and the cross-staff. Also mentioned are the pole and the cord for distance measurement, but no instrument is referred to for the assessment of direction. The geometrical square was used almost exclusively for levelling and for the determination of heights and altitudes, the carpenter's square for horizontal distances, and the cross-staff, again, for fixing heights.

But it was on the Continent that the most important early developments in surveying had taken place. In Nurnberg during the latter part of the fourteenth century, for example, Regiomontanus taught the use of the cross-staff and the geometrical square for everyday practical measurements, while city craftsmen excelled themselves in the construction of compasses. The magnetic compass was clearly used for surveying at this time. It had entered Europe during the twelfth and thirteenth centuries, although the precise date at which it was first used for land surveying, while certainly earlier than the sixteenth century, is unknown. In 1556, for example, Georg Agricola, a German scientist, described the type of compass then used for laying out tunnels and mine-shafts. In Italy, the compass was an adaptation of the mariner's compass, individual positions being fixed by compass-bearing and distance. This is made clear by Niccolo Tartaglia in his *Quesiti et Inventioni* (Venice,

1546), a book containing material gathered much earlier than the date of publication. It was a similar instrument which William Bourne[14] described as his 'Topographical Instrument'. But both Bourne and Digges (whose 'Instrument called Theodolitus' of 1571[15] was basically the same) employed their instruments for triangulation and not merely, as Tartaglia did, for the determination of bearing and distance.

The principle and form of the theodolite was also known quite early on the Continent. Martin Waldseemuller's 'polimetrum' was described in the 1512 Strassburg edition of Gregor Reisch's *Margarita Philosophica*, and the third part of a 1542 manuscript by Jean Roze (Rotz), hydrographer to Henry VIII, also mentions a similar instrument (the 'Cadrant Differential'). But the first person known to have fixed position by intersecting rays was Gemma Phrysius (1509–55) of Louvain,[16] and it was the latter's association with John Dee which led to the introduction of Continental theories into Britain. Dee went to Louvain in 1547 and became a disciple of Phrysius. In 1551 he settled in Mortlake, and his friend Leonard Digges, using Dee's library there, was eventually able to publish his *Tectonicon* in 1556.[17] The first Englishman to describe in print the actual principle of triangulation, however, was Dr William Cunningham of Norwich, who was at Cambridge when John Dee returned from Louvain with the instruments and books of Phrysius, Orontius, and de Rojas. He published his *Cosmographicall Glasse* in 1559, and its derivation from Continental authors is obvious. He describes how to fix the site of a town by two 'angles of position' (Phrysius' phrase), and an instrument graduated 'like a mariner's compass', similar to but not identical with Bourne's. Digges' *Tectonicon*[18] dealt simply with the construction and use of instruments, but his *Pantometria*[19] was designed to teach the correct geometrical principles of surveying and mensuration. In *Pantometria* Digges became the first Englishman to discuss the 'theodolitus', part of the 'intrument topographicall'. This was in essence a form of theodolite, and enabled angles in both the horizontal and vertical planes to be determined.

The rise of geodesy in the seventeenth and eighteenth centuries, and with it the practice of triangulation, created an increased demand for suitable instruments. The three most important were the theodolite, circumferentor, and plane-table, and the several surveying texts which succeeded the appearance of Digges' *Pantometria* demonstrate their increasing importance. Leigh,[20] for example, describes how a surveyor should stand in the centre of every field to be surveyed and look over the whole ground, recording his measurements in a 'paperbooke', but his techniques were essentially those of Digges. Lucar[21] described methods of survey which were based on the plane-table, altitudes being determined by turning the board sideways. Agas,[22] one of the earliest English surveyors to use the plane-table and the theodolite, says how the plane-table, with and without a compass needle, was unsatisfactory for large tracts of land. He found the theodolite preferable for such survey, mentioning an instrument twenty inches in diameter. And in 1616, Rathborne[23] also illustrated the techniques of surveying small areas such as fields.

By the close of the sixteenth century, therefore, the possibilities of the theodolite had already been demonstrated, and the shortcomings of the plane-table were being realised. For the measurement of distance, however, little or nothing had been accomplished, except for variations in the lengths of rods or lines, and the inception of the perambulator or waywiser. Triangulation was increasingly used. In 1653, Leybourn published his *Compleat Surveyor*,[24] and it provides a useful synopsis of techniques which had evolved up to that time. The first part of his book, for example, deals with 'Geometry only'. In the second book 'you have a general description of all the most necessary Instruments used in Surveying, as of the Theodolite, Circumferentor, Plain Table, and the like ...'. Other instruments were the 'cross', described as being suitable for small inclosures of many sides, chains, protractor, and a field-book. In the second book Leybourn writes that the instruments 'in most esteem amongst surveyors' were the theodolite, circumferentor, and plane-table; '... these three ... are sufficient for all occasions ... in laying down of a spacious businesse, I would advise him to use the Circumferentor or Theodolite, and for Townships and small Inclosures the Plain Table'. The third book deals with trigonometry, and the fourth concerns itself with explaining the basic principles of field-survey. In this last book considerable emphasis is placed on triangulation, but its uses are confined to the survey of rather small areas such as fields, and the techniques do not differ appreciably from those described some forty years earlier by Rathborne. The surveyor would, for example, stand as centrally as possible in a field and take bearings to each corner, either by plane-table, theodolite, or circumferentor. Measurements along these bearings could be accomplished by chain. Or intersecting rays could be used to determine the position of various points, the length of the base-line between the two survey positions having been measured by chain.

It is therefore difficult to say precisely which surveying techniques were most likely to have been used by Saxton during the execution of his national survey. Various instruments were available to him, however, particularly for the establishment of horizontal positions (with which his maps were most concerned). The accuracy of such instruments is uncertain however, though clearly they must have left much to be desired. The inception of triangulation towards the close of the sixteenth century implies that he must have sought out elevated vantage points from which to view the territory he wanted to map. But even if he was successful in obtaining assistance and in locating a sufficient number of survey points, the rapidity of the exercise (and the accuracy and consistency which it attained) implies that reference to existing cartographic or documentary evidence probably played no small part in his project. The unfortunate fact about both contemporary field-survey and recorded geographical information is that we can only speculate as to how Saxton applied these resources in various directions. Little has been positively achieved in this field of research. Manley[25] has suggested that Saxton climbed many hills to effect much of his survey. Evans[26] believes that an overestimation of distance was characteristic of much of Saxton's work, possibly implying the use of perambulator or some similar instrument for the assessment of length. These assumptions are undoubtedly relevant so far as they go, but in the last resort we can be definite only with regard to those techniques and instruments available for sixteenth century surveying. How Saxton availed himself of these facilities is likely to remain a matter for reasoned speculation.

Chapter 5

The later editions of Saxton's Atlas

(*a*) *Web's Edition of 1645*
In 1645 William Web published an edition of Saxton's Atlas. Among known copies of Web's atlas are one in the British Library (Map Room; c.7c.3), another in the University Library, Cambridge (class-mark 4.64.3), and a third in the Bodleian Library, Oxford (Gough Maps 96*). The character of the copies varies slightly, although their basic features are sufficiently similar to enable the general format of the atlas to be established, viz: a titlepage (the only introductory leaf) followed by Saxton's maps in set sequence. Web did not appreciably modify the original plates, but some important differences exist between this and Saxton's Atlas. Saxton's introductory pages were omitted, and a titlepage appeared for the first time (Plate 7).

Most of the map titles remain in Latin and in their original cartouches, some are changed into English, while that of Cornwall bears two titles, one the Latin original and the other a new English title. Maps not deriving from Saxton's original plates are occasionally interpolated in the atlas, but their absence from Web's printed index on the titlepage suggests that they were not conceived of as integral parts of the volume. The maps in their correct sequence are listed in Table 6.

Web did not amend Saxton's original plates to any major extent, although the changes he made were pursued with greater vigour by subsequent publishers. His principal modifications were:

(1) a new English title in place of the old Latin title on the maps of 'Anglia ...', Cheshire, Derbyshire, Durham, Lancashire, Lincolnshire, Northumberland and Westmorland;

(2) a new English title for the map of Cornwall, engraved in place of the old dedication to Queen Elizabeth, in the top right of the map. The original Latin title remains;

(3) the Royal Arms and cypher, 'C.R.' (Carolus Rex), of Charles I are substituted for those of Elizabeth on the maps of 'Anglia ...', Cornwall, Durham and Gloucestershire;

(4) the substitution of the Royal Arms of Charles I for those of Elizabeth, but with the cypher 'E.R.' remaining, on the maps of Essex, Glamorganshire and Wiltshire;

(5) the substitution of the arms of Charles I for those of Elizabeth, but without the cypher 'C.R.', on the maps of Cheshire, Denbighshire, Derbyshire, Dorset, Herefordshire, Kent, Lancashire, Lincolnshire, Monmouthshire, Northamptonshire, Northumberland, Shropshire, Somerset, Suffolk, Westmorland and Worcestershire;

(6) the substitution of the date 1642 for the original date on all the maps, except that of Oxfordshire which has the original date crudely altered to 1634; this was almost certainly an error. The original date survives on the maps of Durham, Gloucestershire and Westmorland, despite the addition of the new date;

(7) the division of the map of Yorkshire into two sheets (though not in the Bodleian copy), both of which show the addition of inset plans ('The Citye Of Yorke' on the plate of western Yorkshire, and "The Towne of Hvll" on the plate of eastern Yorkshire). The addition of an inset plan ('Barwick') also occurs on the map of Northumberland. The insertion of these insets involved the erasure of certain of the detail from Saxton's original

THE MAPS OF ALL THE SHIRES IN England, and Wales.

EXACTLY TAKEN AND TRVLY

Decribed by

CHRISTOPHER SAXTON.

And graven at the Charges of a private Gentleman for the publicke good.

Now newly Revised, Amended, and reprinted.

1 The Generall Map of England.

2 Cornwall.
3 Devonshyre
4 Somersetshyre
5 Dorsetshyre
6 Hampshyre
7 Sussex Surry Kent and Midlesex
8 Essex
9 Suffolke
10 Norfolke

11 Cambridge Huntington Rutland Northampton & Bedford
12 Hartfordshyre
13 Buckinghamshyre Oxfordshyre and Berkshyre
14 Wiltshyre
15 Glostershyre
16 Herefordshyre
17 Shropshyre
18 Worsetershyre

19 Warwick & Lecester
20 Lincolne & Noitingham
21 Darbyshyre
22 Staffordshyre
23 Cheshyre
24 Lancashyre
25 and 26 Yorkshyre
27 Durham
28 Cumberland & Westmerland
29 Northumbrland.

The shires in Wales.

30 Monmouthshyre
31 Pembrokeshyre
32 Glamorganshyre
33 Radnor Brecknock Cardigan and Carmarthinshyre
34 Mongomrye & Merionith
35 Anglesey & Carnarvan
36 Denbigh & Flintshyre

Printed for Wlliam Web at the Globe in Cornehill. 1645.

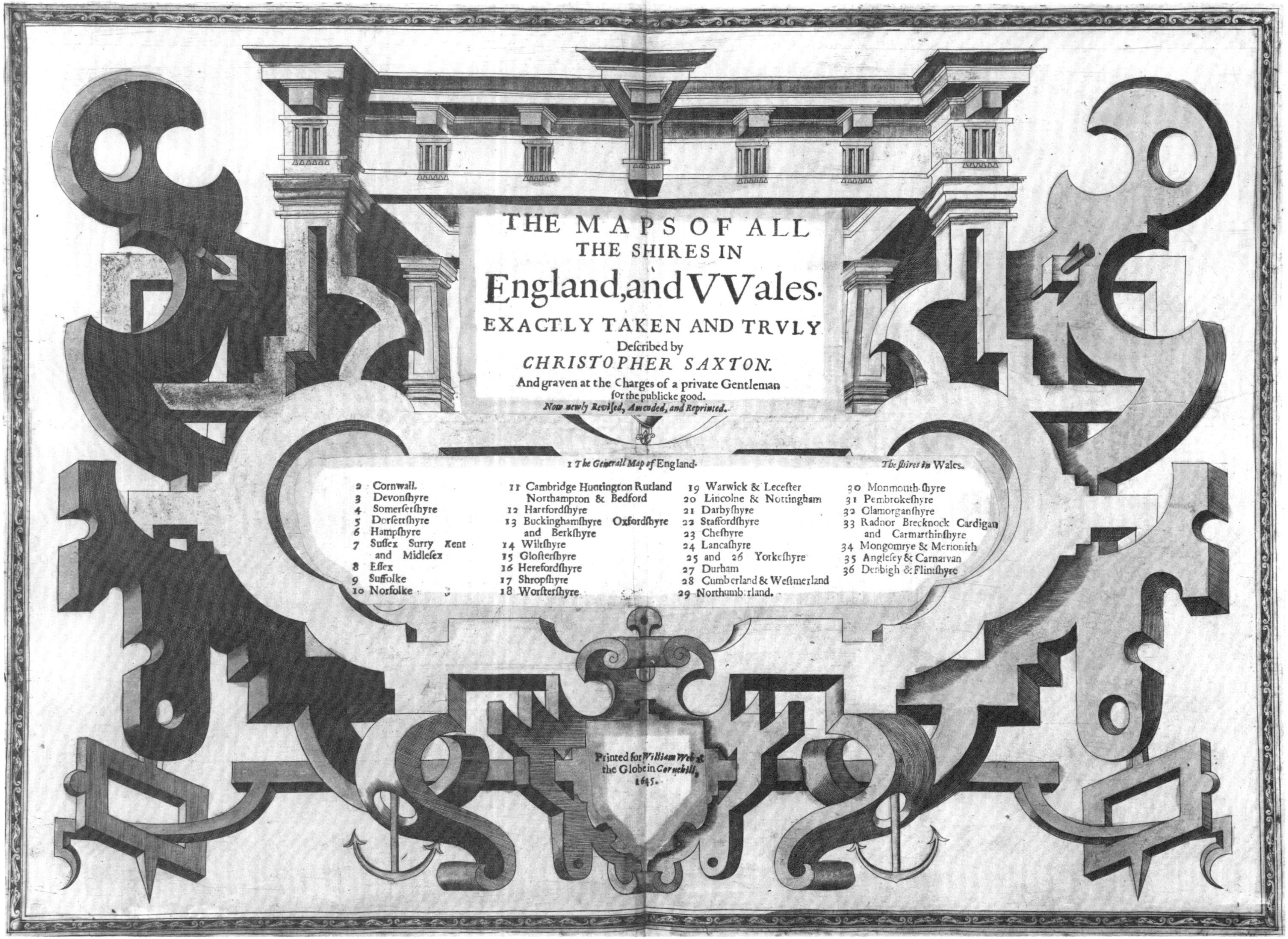

Titlepage to William Web's edition of Saxton's Atlas, 1645.

PLATE 7

plates, e.g. on the map of Yorkshire the inset of Hull replaces the original title cartouche and the Royal Arms, and that of York replaces Seckford's arms;

(8) the addition of the divisions of the wapentakes and their names to the map of Lincolnshire, and of the wapentakes (without names) to the map of Yorkshire.

(b) Whitaker's Projected Edition of 1665

Lynam[1] referred to the possibility of an edition of Saxton's Atlas having appeared between the Web publication of 1645 and that of Philip Lea in *c.*1689. To support his hypothesis he remarked that there had appeared on several of Lea's plates of *c.*1689 the date 1665, or palpable alterations of it, and that the royal cypher 'C.R.', already engraved on four maps by Web, had been added onto several maps by *c.*1689; this implied that such maps were issued as early as 1685, when James II succeeded Charles II to the throne. These suggestions stimulated the interest of Harold Whitaker, a collector of old maps, who compared several editions of Saxton's Atlas, particularly the two Lea editions (*c.*1689 and *c.*1693) and that of Web.[2] His conclusions were basically as follows:

(1) A few of the maps in Lea's edition of *c.*1689 contained the date 1665 or alterations of it; even where this date does not appear, some plates reveal features which suggest a state earlier than *c.*1689:

   (a) Herefordshire is dated 1665 (Table 7, No.13 and Appendix 13, No.14).

   (b) Somerset and Wiltshire have the date 1689, but they are clearly alterations from 1665 (Appendix 13, Nos.28 and 33). Chubb[3] had already recorded the earlier state of the Somerset map, but erroneously dated it 1685.

   (c) Norfolk shows an imperfectly erased 1665 (Appendix 13, No.21).

   (d) Shropshire shows faint traces of what might have been two sixes and a distinct trace of a figure five, all in appropriate positions for a date 1665 (Appendix 13, No.27).

   (e) Northamptonshire has a new cartouche and title, placed in the lower right-hand corner. Beneath this there are traces of an erased inscription (Appendix 13, No.22) which must have been made after Web's edition (when this particular space was blank); but it is now undecipherable.

   (f) A copy of *c.*1693 in the Manchester Public Library ('The Shires of England And Wales ... by Philip Lea'; B.R. F 912 42 S6) has a map of Suffolk with the title in a different cartouche from that used by Saxton and Web, showing clear signs of a date 16(?)5 having been erased. Its title lacks the words 'By C.S. Corrected and Amended by P. Lea' which occur on the map in the atlas of *c.*1689. The map of Lancashire in the Manchester copy is similarly earlier in character to that in the edition of *c.*1689, its title lacking the words 'Discribed by P. Lea'. Both these copies of the Suffolk and Lancashire maps lack the crosses and crowns on the symbols of the towns, although in other respects they show the additions formerly attributed to Lea in *c.*1689 and *c.*1693 (Table 8, Nos.29 and 16).

   (g) A copy of the edition of *c.*1693 ('The Shires of England and Wales'; University Library, Cambridge: Atlas. 4.69.2) has its map of Staffordshire replaced by one which is earlier than that in the edition of *c.*1689. Its title is still in Latin, but Web's date is erased and the space left blank. It also lacks the crosses, crowns, and mitres, but in other respects has the additions formerly attributed to Lea (Table 8, No.28).

(2) The Royal Geographical Society's copy of *c.*1689 ('All the Shires of England and Wales ... by Philip Lea; 264.H.17) shows the addition of reference letters to sixteen of the

maps (Appendix 12) along the top and down the left side, i.e. Cornwall, Derbyshire, Dorset, Durham, Essex, Hampshire, Herefordshire, Lancashire, Monmouthshire, Norfolk, Shropshire, Somerset, Staffordshire, Suffolk, Wiltshire, and Worcestershire. If this was an innovation of Lea's in *c.*1689, it is conspicuously evident that he did not extend it to those maps not yet possessing it in his edition of *c.*1693. It was a modification which he was not anxious to continue, and therefore quite likely not responsible for in the first place.

(3) Lea's edition of *c.*1689 shows the addition of extra place-names, hills, and rivers, more especially in the adjacent counties, to fifteen of Saxton's maps (Derbyshire, Dorset, Durham, Gloucestershire, Hampshire, Herefordshire, Lancashire, Monmouthshire, Norfolk, Northamptonshire, Shropshire, Somerset, Staffordshire, Wiltshire, Worcestershire) and to two copies (Devon and Northumberland). The innovation was not extended into his edition of *c.*1693, save for Cheshire and Shropshire, and was therefore again more than likely initiated by some other person.

(4) Web had added the royal cypher 'C.R.' to three of his county maps, i.e. Cornwall, Durham, and Gloucestershire. The edition of *c.*1689 shows it added to eight others, i.e. Derbyshire, Glamorganshire, Herefordshire, Lancashire, Monmouthshire, Northamptonshire, Staffordshire, and Wiltshire (Appendix 12). Yet Lea would not have inserted such a cypher after 1685.

These facts prompted Whitaker to suggest that an edition of Saxton's Atlas was published in 1665. It has not survived, and its contents and character must be surmised. Yet the suggestion is plausible. The edition of *c.*1689 lacks Saxton's plates of Devon and Northumberland, which never subsequently reappeared; it is possible that they were destroyed in the Great Fire of London, 1666, and it is similarly possible that any edition just published might also have been destroyed.

On the basis of the criteria examined above it is possible to assemble a list of maps which probably belonged to an edition of 1665, the characteristics of which can be generally defined. Those county maps which appeared to meet the format of such an edition were Cheshire, Cornwall, Derbyshire, Dorset, Durham, Essex, Glamorganshire, Gloucestershire, Hampshire, Herefordshire, Lancashire, Monmouthshire, Norfolk, Northamptonshire, Shropshire, Somerset, Staffordshire, Suffolk, Wiltshire and Worcestershire. These twenty county maps constituted, with but few modifications, Whitaker's edition of 1665. Essex was omitted, however, because Saxton's plate had been lost for the edition of *c.*1689 (though it reappeared, with reference letters, in *c.*1693). The map of Cheshire was also omitted, the addition of extra place-names (the map's only qualification) being too tenuous a criterion. Whitaker adds the map of Yorkshire to the above list, but his reasons for doing so are unclear and seemingly unjustified. In his writings subsequent to 1938, Whitaker includes the projected editions of 1665 among those later editions of Saxton's Atlas to which he referred,[4] though prior to that date the edition is unmentioned.[5] He considered it extremely likely that all Saxton's maps had undergone revision by an unknown cartographer in *c.*1665, although the evidence for this was insufficiently strong to justify their inclusion in his list of 1939; for example he mentions the possibility of maps of Cheshire and Northumberland having been published at this time, despite their omission from his projected atlas.[6]

The characteristics of the maps in this edition can only be assumed in general form. Where a map possessed details added by Web in 1645 which remained intact in *c.*1689, then the edition of 1665 must have possessed them also. Further, those modifications which seem to have been uncharacteristic of Lea, but which nevertheless appeared in his edition of *c.*1689 (reference letters and extra names), were presumably added in 1665. Features such as the date 1665 and the royal cypher 'C.R.' also occurred where evidence proves this to be the case. Such additions and substitutions would of necessity incur some amendment to the format of the maps concerned.

All Saxton's original titles were in Latin. Web changed eight of these (excluding 'Anglia ...') into English (Cheshire, Cornwall, Derbyshire, Durham, Lancashire, Lincolnshire, Northumberland, and Westmorland), the rest remaining in Latin. By c.1689 all the others (with the exception of Yorkshire which was without a title) were in English. At some time between 1645 and c. 1689 a linguistic change in the titles must have occurred, and one can presume these up to a point by selecting the common denominator from both editions (compare Tables 6 and 7). The Yorkshire map had had its title removed by Web, and presumably the unknown cartographer was engaged in a process of converting this and all other titles into English (though in the case of Staffordshire and Yorkshire it seems the task was not completed). The list of map titles was therefore possibly as follows:

1. Cornwal / with ye severall hundre / ds, truly described. (*Continuity of title, 1645 to c.1689.*)
2. An Exact Map of Darbieshire. (*Continuity of title, 1645 to c.1689.*)
3. A true description / of / Dorsetshire. (*Assumption based on edition of c.1689.*)
4. The County Palatine / and Bishoppricke of / Durham. (*Assumption based on edition of c.1689.*)
5. Hampshire. (*Assumption based on edition of c.1689.*)
6. Glamorgā / Shire. (*Assumption based on edition of c.1689.*)
7. Glocester-Shire. (*Assumption based on edition of c.1689.*)
8. The County of / Hereford / resurveyed / and / enlarged. / An^o: 1665. (*Assumption based on edition of c.1689.*)
9. The / County Pallatine / of / Lancaster. (*Positive evidence from edition of c.1693, Manchester Public Library: B.R. F 912 42 S6.*)
10. Monmovthshire / heretofore part / of Wales, / But now added / to other English / Shires in ye Con: stant Circuite / of the Iudges. (*Assumption based on edition of c.1689.*)
11. The / Covnty of / Norfolke / exactly plotted ... 1665. (*Assumption based on edition of c.1689.*)
12. The County of / Northampton, / togeather wth ye three small / Counties of / Bedford Hvntingdon / & Rvtland. (*Assumption based on edition of c.1689.*)
13. Shropshire / accurately drawen / and sett forth. Anno 1665. (*Assumption based on edition of c.1689.*)
14. The County of / Somerset / exactly plotted / Published An^o 1665. (*Assumption based on edition of c.1689.*)
15. Staffordiae / Comitatu' pfecte et / absolute elaboratu haec / tibi tabula exhibet. (*Positive evidence from edition of c.1693, Cambridge University Library; Atlas. 4.69.2.*)
16. Svffolke / described. Anno 1665. (*Positive evidence from edition of c.1693, Manchester Public Library; B.R. F 912 42 S6.*)
17. Wiltshire wth Salisbvry Citty / & Stone heng described An^o. 1665. (*Assumption based on edition of c.1689.*)
18. Worcestershire / and Citty / exactly described. (*Assumption based on edition of c.1689.*)
19. YORKSHIRE. No title. (*Continuity, 1645–c.1689.*)

It may be presumed that the unknown publisher continued the practice, already begun by Web, of adding hundreds to the maps, as well as town plans. Crosses, crowns and mitres, which appear on all the maps in the edition of c.1689, are almost certainly Lea's additions; they show uniformity from map to map, and appear to be an innovation pertaining to the whole set of county maps, not merely the extension of a practice which had been applied to only nineteen maps at a previous date. Shields, already in vogue, may be presumed to have occurred frequently.

The nature of introductory pages to the edition of 1665 can, like that of the maps, only be inferred indirectly. Mr John Gardner once possessed a copy of Lea's 'All the Shires...', c.1689 (this Lea-version of the atlas has been acquired by the British Library), in which the maps were identical with those in the Royal Geographical Society's copy (264.H.17) but in which the titlepage was in an earlier version. There was neither the sub-heading 'Wales' before the list of Welsh counties nor an 'Explanation' of the symbols, both of which occur in the 'Table of the Shires'. It is possible that this was the impression of the titlepage for the 1665 edition, although Whitaker preferred to believe it formed merely an impression from an unfinished plate (possibly by Lea), the spaces being left clear for the inclusion of the items in question. There exists no evidence to support any assumptions regarding further introductory pages in this edition.

(c) *The Editions of Philip Lea*

It appears that at some date between 1665 and 1689, the instrument-maker and mapseller Philip Lea acquired Saxton's plates in their 1665 state. He embarked on a major modification of the plates, which he completed with the publication of an edition of the atlas c.1693. c.1689, however, he also issued an interim edition, representing a stage in the evolution of the plates; some had hardly been altered at all, while a few had almost reached the stage of sophistication which others only aspired to c.1693. Lea also issued the edition of c.1693 as a French publication, although this constituted essentially the same work as its English counterpart.

Lea's three editions were entitled as follows:

(1) c.1689. All / the Shires of / England / and / Wales / Described by / Christopher Saxton / Being the Best and Original Mapps / With many Additions and Corrections by Philip Lea.

(2) c.1693. The Shires of / England / And / Wales / Described by / Christopher Saxton / Being the Best and Original Mapps / With many Additions and Corrections / Viz: ye Hund^{ds}, Roads, etc. by Philip Lea. Also the New Surveis / of Ogilby. Seller. etc.

(3) c.1693 (French). Atlas / Anglois / Contenant / Les cartes Nouvelles & tres Exactes / Des provinces, duchés, Comtes, & Baronies / du Royaume / d'Angleterre. / le tout Enrichi des Plans des Villes & des Armes / de la Noblesse. / London, / Par P. Lea.

Recognition of two Lea editions has been a process of investigation rather than spontaneous discovery. It has already been remarked (Chapter 5b) that Chubb failed to distinguish the two.[7] As late as 1927 he was still grouping both under the date 1690.[8] This is a little surprising, because Fordham had on several occasions already referred to the existence of Saxton maps of c.1689,[9] and Curwen[10] had mentioned maps of Cumberland and Westmorland which he dated at 1689–90 (though he was erroneously referring here, in the event, to Lea's edition of c.1693). It was left to Whiticker[11] to identify two distinct editions of Lea's map of Yorkshire.

The dates of these three editions are based on a rather limited fund of evidence. In the case of the edition of c.1689 no indication as to the date of publication occurs on any of the introductory leaves, and it is from cartographic testimony alone that one can hazard an assumption. The criteria which suggest 1689 as the likely date of publication are:

(1) the date 1689 occurs on the maps of Somerset and Wiltshire, though on no others;
(2) Ogilby's *Britannia*, from which the roads on some of the maps were copied, e.g.

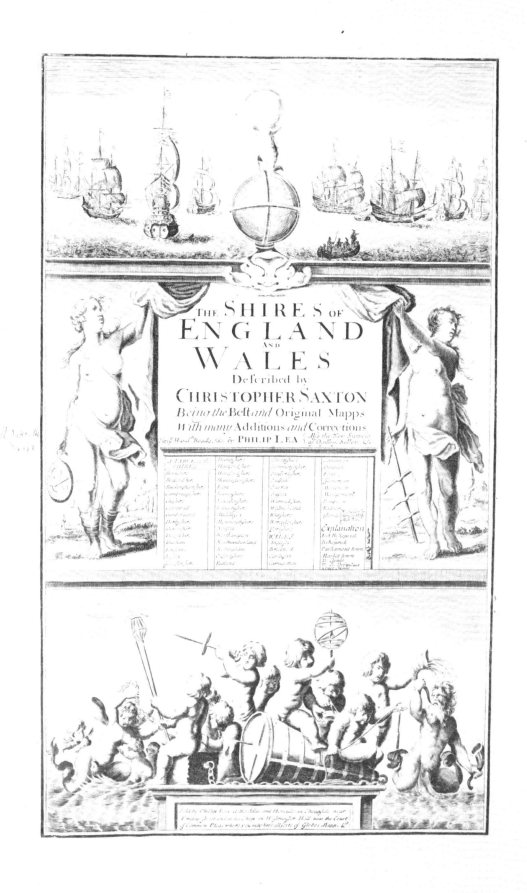

Titlepage to Philip Lea's edition of Saxton's Atlas, c.1693. The titlepage to Lea's edition of c.1689 is identical in all but a few respects.

Yorkshire, and from which the map of Essex was taken, appeared in 1675, so that Lea's interim edition clearly post-dated this by at least several years;

(3) the royal initials 'W.R.' (William III, r.1689–1702) replace 'C.R.' on the map of 'Anglia ...', though on no others, indicating a date not earlier than 1689;

(4) the use, for the map of the 'Islands', of Greenville Collins' chart of the Scilly Isles, the earlier state of which is dated 1689.

No definite date of publication for this edition can therefore be postulated, but a date c.1689 seems plausible.

Similarly with the edition of c.1693, it is impossible to be precise; indeed, the evidence available for examination is more meagre than that for the interim edition. Whitaker[12] examined the British Library's copy (Map Room; c.7c.4) and observed that Lea's map of Kent (not based on an original Saxton plate) is dedicated to John Tillotson as Archbishop of Canterbury from 1691–4 (Table 8, No.15), so that the map probably appeared during this short span of time. In addition, Whitaker's own copy (University of Leeds, Brotherton Library; Whitaker Coll.2) had the ownership inscription of Narcissus Luttrell, with the date of acquisition as c.1693. In the absence of further evidence this has been assumed to be a likely date of publication. The French edition is given the same date on the grounds that it is essentially the same atlas as its English counterpart, the only major difference relating to the French titlepage. It was probably prepared at the same time as the English edition, but with a particular market in mind.

In all three editions there was no established sequence of maps (Tables 7 and 8). This is indicated by two criteria: (a) there occurs, on the titlepage of each of the two English editions, 'A Table of the Shires'. This is a list of all the counties arranged alphabetically, with an empty space alongside the list in which any owner of a copy could complete in manuscript the pages of the atlas on which the map of a particular county occurred in his own copy (Plate 8). The format of this 'Table' indicates that there was no set sequence of maps; (b) in extant copies the maps vary with regard to sequence. Table 7 lists maps appearing in the edition of c.1689, and Table 8 those in the edition of c.1693.

Introductory pages to the English editions vary. The first page of both consists of the titlepage, which with but a few variations is identical in each case (Plate 8). It is a reverse copy of the titlepage in Hendrik Doncker's *The Sea-atlas or the Watter-World... (Amsterdam, 1660)*. The title occurs in a decorated cartouche in the centre of the page. Below this is the 'Table of the Shires', incorporating an 'Explanation' of those symbols which Lea added to the maps. In both editions these include symbols for each of an 'Arch. Bishoprick', 'Bishoprick', 'Parliament town', and 'Market town', but the edition of c.1693 has a further three symbols not appearing in the earlier edition, i.e. for 'D. Roads', 'D.R. Dependent', and 'Cross Roads'. Lea's imprint occurs at the foot of the titlepage in both editions.

The titlepage constitutes the only introductory page in the edition of c.1693, and is less embellished in the French version. In this the type is set up in roman, and the only decorative feature is an engraving of a male figure holding up two bowls of fruit and surrounded on either side by a dog. In the edition of c.1689 the titlepage is the first of three introductory pages. The second single page consists of a type-printed table headed 'The Cities and Principal Towns in England / and Wales, Alphabetically Digested, and their Computed / and Measured Distance from the Standard in Cornhil, / London; with the Degrees of Longitude and Latitude'. Then follows a double-page engraving of the arms of Charles I. No set sequence appears to have existed for these introductory pages.

Lea's modifications to Saxton's plates were effected steadily over a period of years, and reached their culmination c.1693 with the appearance of his final edition. The edition of *c*.1689 merely represents a stage in this process, by which time some plates had been considerably more modified than others. The changes which he made were considerable, and may be briefly summarised as follows:

(1) The English rendering of map titles and scales where this had not already been effected. This modification was complete by *c*.1689.

(2) The addition of crowns, crosses and mitres to represent different categories of town. These were explained in his introductory 'Table of the Shires'. The edition of *c*.1689 shows the addition of crosses to denote market towns on all the county maps. Crowns to denote parliamentary boroughs and mitres to denote bishoprics are also shown, though far less exhaustively. The edition of *c*.1693 shows this innovation completed.

(3) The addition of roads. This process was begun in the edition of *c*.1689, but no explanation of the symbols occurred until they were added to the 'Table of the Shires' in *c*.1693. The roads, copied from Ogilby's survey,[13] were shown fully on six maps in *c*.1689 (Gloucestershire, Hampshire, Lancashire, Northumberland, Worcestershire and Yorkshire) and partially on one (Westmorland). The edition of *c*.1693 shows all the maps completed in this respect.

(4) The erasure of Seckford's arms and occasionally of the Royal Arms, where this had not already been effected in 1665. Seckford's arms survived only on the maps of Kent and Westmorland in *c*.1689, though twenty-one maps retained the Royal Arms.

(5) The insertion of hundreds, wards and wapentakes, to continue the process begun in earlier editions. The edition of *c*.1689 shows the Yorkshire map completed in respect of hundred-names, and the addition of hundreds on a further nine maps (Anglesey, Cheshire, Devon, Kent, Montgomeryshire, Oxfordshire, Pembrokeshire, Radnorshire and Warwickshire). The edition of *c*.1693 shows the hundreds on the remaining four (Denbighshire, Durham, Northumberland and Westmorland).

(6) The addition of town plans and of heraldic shields where these had not been added previously. This modification was gradually extended through both editions, yet on the Yorkshire map in the edition of *c*.1693 the plan of York was erased.

(7) The occasional insertion of features of peculiar local interest, e.g. 'Offa's Ditch' on the map of Denbighshire.

(8) The alteration of the date to 1689 on two maps (Somerset and Wiltshire) in the supposed edition of that year, and the addition of Lea's imprint to all the maps in both editions.

(9) The further addition of a few extra names on the maps of Cheshire and Shropshire.

These numerous changes to the maps indirectly resulted in two further modifications: (a) the appearance of several empty spaces on the maps in the edition of *c*.1689, where erasures had not as yet been replaced by additional detail; these had for the most part been filled in by *c*.1693, and (b) the rearrangement of the format of the maps, particularly by *c*.1693, a feature more extensive in some than in others.

Lea's edition of *c*.1693 was cartographically the final state of Saxton's maps (Appendix 13). After this, only the imprints were altered, geographical and cartographic features remaining unaffected.

(d) The final editions of Saxton's Atlas

The first edition to appear subsequent to Lea's was that of George Willdey, *c*.1730[14] (*The Shires of England and Wales ... Sold by Geo: Willdey at the Great Toy, Spectacle, China-ware and Print Shop, at the Corner of Ludgate Street near S*^{t}*. Pauls London*). Willdey's imprint is added to each map, but otherwise they remain unchanged. The atlas was evidently conceived of as a travelling atlas, being bound in a limp cover with all the maps cut down to their

Saxton's Atlas, 1579. Map of Northamptonshire, etc. State II.

PLATE 9

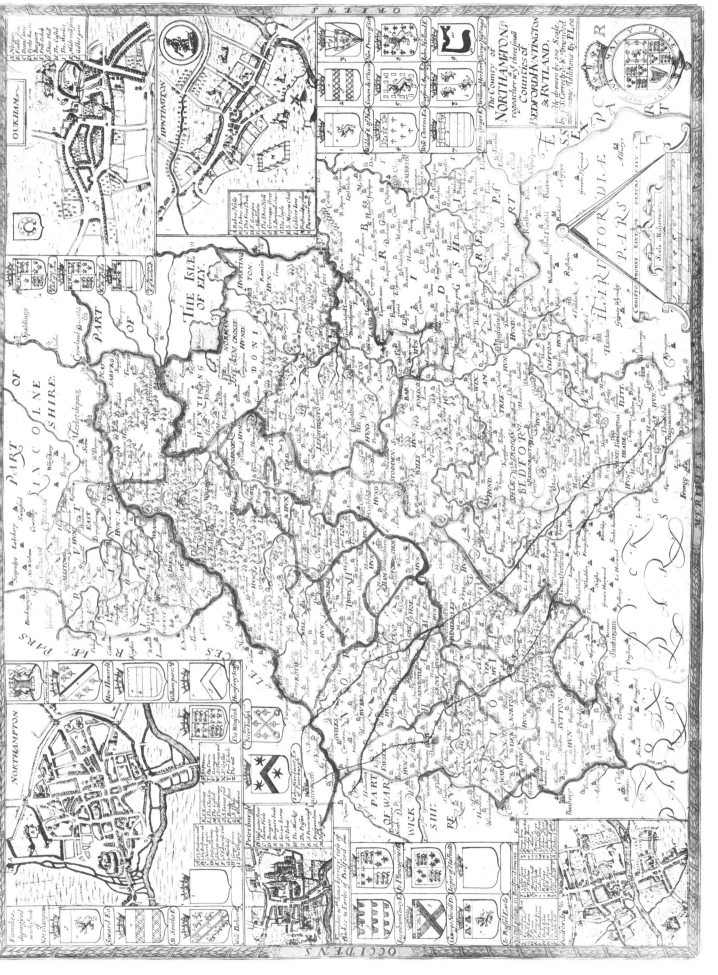

Philip Lea's edition of Saxton's Atlas, c.1689. Map of Northamptonshire, etc.

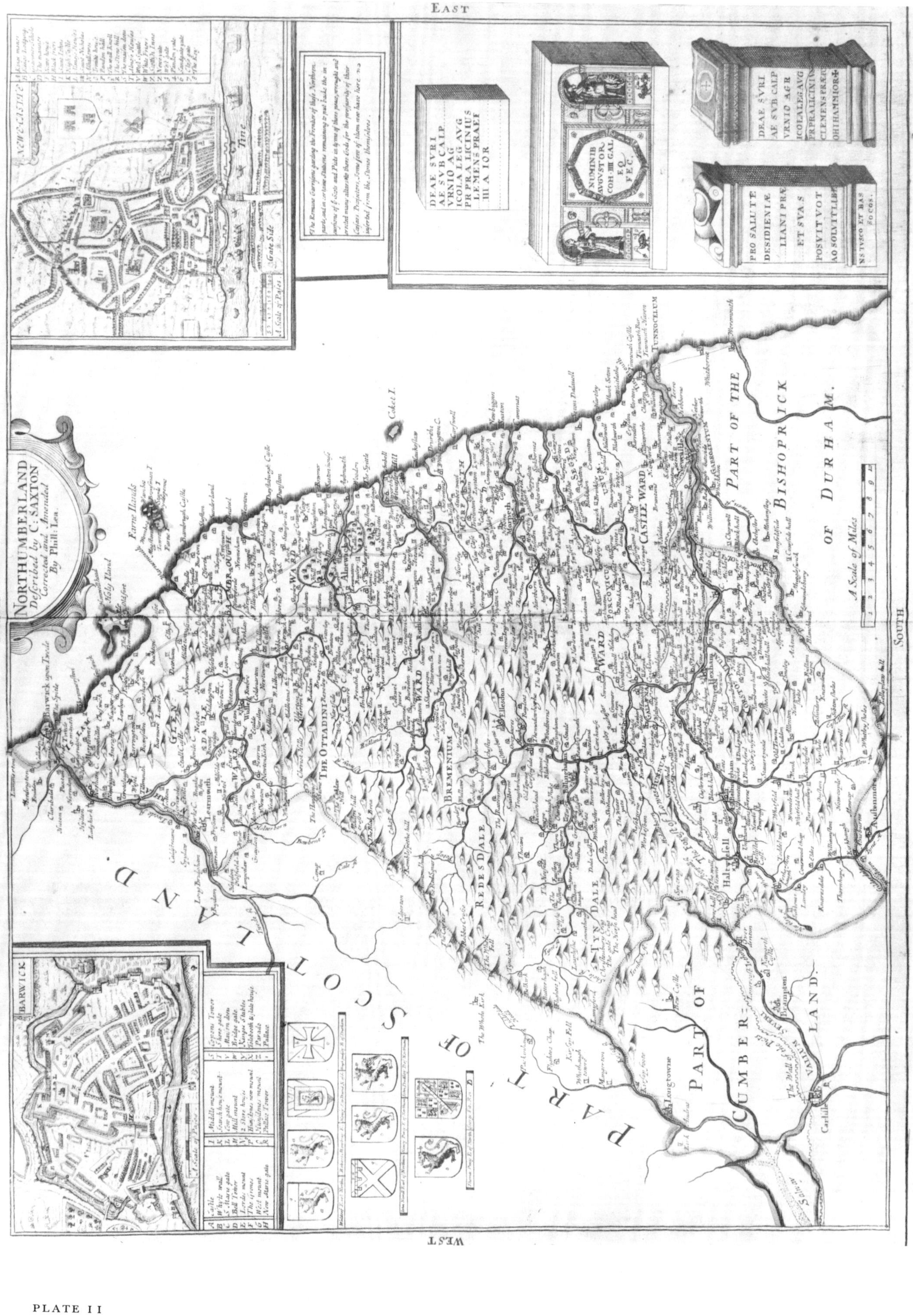

PLATE II

Philip Lea's edition of Saxton's Atlas, c.1689. Map of Northumberland.

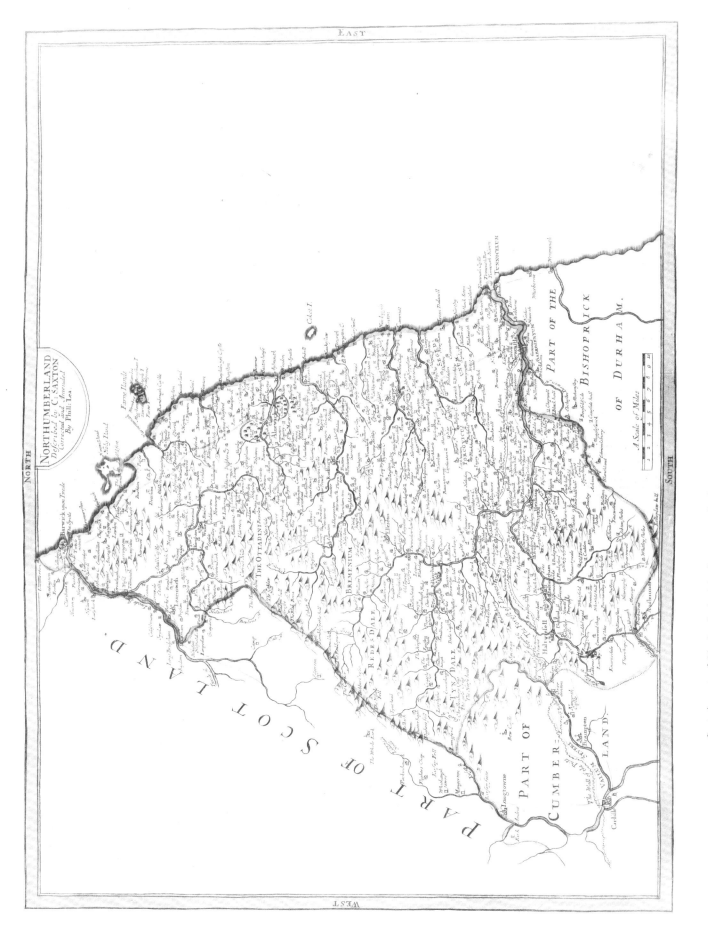

Saxton's map of Northumberland in the process of subsequent modification. The Royal Arms, English title, Seckford's arms, and the line-scale, all of which had appeared on William Web's map (1642), have been erased. The detail which was eventually to appear on Lea's plate (see Plate 11) has not yet been engraved.

PLATE 12

borders (some of Willdey's imprints having been lost in the process). Another edition by, Thomas Jefferys, was published c.1749 (*The Shires of England and Wales . . . Sold by Thomas Jefferys, Geographer to his Royal Highness the Prince of Wales; in Red Lyon Street near S^t. John's Gate*). Willdey's imprints have been erased, leaving many of the maps scarcely distinguishable from those of Lea's last edition in c.1693. Saxton's Atlas finally appeared in c.1770, when the plates were acquired and the maps issued by C. Dicey & Co., again without significant modification. There is no titlepage to this edition, so it is either lost or was no longer used. The plates are so worn that the formerly imperfectly erased Willdey impressions have completely disappeared, and the large map of Yorkshire has been replaced by Sutton Nicholl's map of 1711.

(*e*) *Conclusion* (Extant copies of later editions of Saxton's Atlas are given in Appendix 14). All Saxton's plates had reached their final state, cartographically, by c.1693; alterations thereafter were minimal. Very few reveal any changes during the period of publication of the initial edition in 1579. Only the maps of 'Anglia . . .', Norfolk and Northamptonshire witnessed modifications, and these (described in Chapter 3) were very minor (Plate 9).

Not all maps were drastically modified in Web's publication of 1645, though some experienced more changes than others. The map of Cheshire, for example, acquired amongst other things an English title, whereas on the map of Northamptonshire only the Royal Arms and dates were altered. Evidence has not been forthcoming to justify the inclusion of many maps in an edition of 1665. That of Northamptonshire, however, probably did appear during this year, its cartographic characteristics being those which it possessed in Lea's edition of c.1689 (Plate 10), after which no major alterations were effected.

Only occasionally does one discover a map representing an intermediate state. The Bodleian copy of c.1689 (Douce Prints b.28), however, possesses a map of Northumberland which is intermediate between the editions of Web (1645) and Lea (c.1689). Lea's edition (Plate 11) witnessed the substitution of Roman altars and a plan of Newcastle for, on Web's plate, the Royal Arms, English title, Seckford's arms, and line-scale; heraldic shields had also been added beneath Web's plan of Berwick. The Bodleian map (Plate 12) shows Web's detail erased, but the spaces still blank; Lea had evidently not yet added his new information to the plate.

TABLE 6
Web's edition, 1645. Titles of maps

(British Library, Map Room; c.7c.3)

All maps in this copy, including those not deriving from Saxton's plates (distinguished by an asterisk), are included. Full titles are detailed only in the case of (a) changes in the titles of Saxton's plates, and (b) other maps interpolated in the Atlas.

1. Anglia / The Kingdome Of / England / and Principality of / Wales / exactly Described / An⁰ Dñi / 1642.
2. Promontorivm Hoc / In Mare Proiectum / Cornvbia Dicitvr. *Also:* Cornwal / with y^e severall hundre / ds, truly described / 1642.
3. DEVONSHIRE.
4. SOMERSET.
5. DORSET.
6. HAMPSHIRE.
7. KENT.
8.* Provinciae / Cantii / Vulgo / Kendt / Nova Descriptio. Scala Milliaria Anglica 6 (=59 millimetres).
9. ESSEX.
10.* Essexiae / Descriptio. / The Description of / Essex. / Amstelodami, / Sumptibus Ioannis Ianssonii. Milliaria Anglica 9 (=108 millimetres). Milliaria Germanica Communia, 2 (=75 millimetres).
11. SUFFOLK.
12. NORFOLK.
13. NORTHAMPTONSHIRE.
14. HERTFORDSHIRE.
15. OXFORDSHIRE.
16. WILTSHIRE.
17. GLOUCESTERSHIRE.
18. HEREFORDSHIRE.
19. SHROPSHIRE.
20. WORCESTERSHIRE.
21. WARWICKSHIRE.
22. Lincolne & Notinghamshire / With theire severall Hundreds / and Wapontakes most Exactly / drawne and described. / Anno Domini. 1642.
23. An / Exact Map of / Darbieshire / Anno: 1.6.4.2.
24. STAFFORDSHIRE.
25. The County / Palatine of / Chester.
26.* The / Covntye Palatine / Of / Chester / Comitatus / Cestrensis. Amsterdami Apud Ioannem Ianssonium. The Scale of Miles. 5 (=54 millimetres).

27. The / County Pallatine / of / Lancaster / Anno. 1.6.4.2.

28. YORKSHIRE, WEST (no title).

29. YORKSHIRE, EAST (no title).

30.* Provincia / Eboracen = / sis / Yorke-Shire / Amstelodami, apud Ioan: Ianssonium. Milliaria Anglica. 18 (=94 millimetres). Milliaria Germanica communia 4 (=83 millimetres).

31. The / County Palatine / of / Dvrisme / Exactly drawne / 1642.

32. Comberland / and / Westmorland / Exactly described / 1642.

33. Northvmberland / Truely Drawne & / Described. / Anno: 1.6.4.2.

34. MONMOUTHSHIRE.

35. PEMBROKESHIRE.

36. GLAMORGANSHIRE.

37. RADNORSHIRE.

38.* Ceretica / Sive / Cardiganensis / Comitatvs / Cardigan / Schire. 6 (=64 millimetres).

39. MONTGOMERYSHIRE.

40. ANGLESEY.

41. DENBIGHSHIRE.

42.* Carte / Generalle de la Grande / Bretagne, jadis / Albion, / Et du Royaume d'Irelande, auec / les Isles circonuoisines / despandantes desdits / Royaumes; / auec plusieurs obseruations / 1644,. A Paris chez Iean Boisseau.

43.* Carte / Generale Des / Royavme D'Angleterre / Escosse Et Irlande / Auecq les Isles circonuoisines / Conues toutes soubs le nom de / Britanniqves / Nouuellement dressee & tiree de Camb: / dene spede & autres Par N. Sanson / geogre Ordre du Roy.

44.* Pascaart / Vaut / Canaal / Tusschen Engelant en Vrancryck, alsmerde ge / heel serlanten Schotlant, waer in men claer can / sien de rechte distantien en courssen tusschen alle / havens en droogtē, alles op syn ware hoogten geleyt. / Tabula Hydrographica / Freti seu Canalis / quod est / Inter Angliam et Galliam, nec non Hiberniam et / Scotiam, in quā longitudines latitudines .q omnium / Postuum breviumque, in sua cuiusque poli Eleva; / tione graphicē depinguntur novissimē discripta. / Amstelodami a Ioanne Ianssonio.

45.* *Two-page description of 'The Fenns', with a map interpolated between the two pages:* A general Plott / and description of / the Fennes and sur = / ounded grounds in / the sixe Counties of / Norfolke, Suffolke, / Cambridge, with in / the Isle of Ely, Hun = / tington, Northamp = / ton and Lincolne etc. Amstelodami Sumptibus Henrici Hondii. 1632.

46.* Middelsexiae cum / Hertfordiae / Comitatu: / Midlesex & Hertford / Shire. Amstelodami Apud joannem janssonium.

47.* Anciens Royaumes / De Kent, D'Essex, Et De Sussex: / ou sont Aujourdhuy les Comtés / De Kent, D'Essex, Middlessex, Et / Hartford De Sussex, Et Surrey. / Avecq les Pas de Calais, et / Partie des Costes du Pays Bas, / de Picardie, et Normandie. / Par le S. 'Sanson d'Abbeville Geogr. ord. du Roy. / Avecq. Privilege pour Vingt Ans. / A Paris. / Chez l'Auteur. / Pres de S. Germain l'Auxerrois. / 1654.

48.* Provinces d'West; / autrefois / Royaume d'Westsex: / ou sont Aujourdhuy les Comtés / Hant-Shire, et l'Isle de Wight, Barck-Sh. / Wilt-Sh. Dorcet-Sh. Somerset-Sh. Devon-Sh. et Cornwall, etc. / Par le S. 'Sanson Geogr. Ord. du Roy. Avecq Privilege pour Vingt Ans. / A Paris. / Chez l'Auteur. / 1654.

49.* Anciens Royaumes De / Mercie et East-Angles: / ou sont les Comtés, ou Shiries de / Chester, Darby, Nottingham, / Lincolne, Rutland, Leicester, / Stafford, Shrop-Sh, Hereford, / Worcester, Warwick, Northampton, / Huntington, Bedford, Buckingham, / Oxford, et Glocester / en Mercie: / Cambridge, Norfolk, et Suffolk / en East-Angles. / Par le S. 'Sanson Geogr. ord. du Roy. / 1654.

TABLE 7

Maps in the Saxton Atlas by Lea, c.1689

Sequence as in the Royal Geographical Society's copy (264.H.17). Maps which do not derive from Saxton's plates are indicated by an asterisk. Maps belonging to the edition, but missing from the Royal Geographical Society's copy, are appended at the foot of the table in no particular sequence.

1.* A New Map / Of England / And Wales With / The Direct and Cros Roads / Also the number of Miles between the / Townes on the Roads by inspection in / figures. Sold by Phillip Lea Globemaker / at the Atlas and Hercules in Cheap / side near Fryday / Street.

2. Anglia. / The Kingdome of / England / and Principality of / Wales / exactly Described. / P. Lea excudit.

3.* Northumberland / Described by C: Saxton. / Corrected and Amended / by Phill: Lea.

4. The County Palatine / and Bishoppricke of / Durham, / Discribed by C.S.: Corrected and / Amended with many Additions / By Phil: Lea.

5. Comberland / and / Westmorland / exactly described.

6. The / County Pallatine / of / Lancaster / Discribed by P. Lea.

7. YORKSHIRE (no title).

8. Lincolne Shire / and / Nottinghame Shire / By C. Saxton.

9. An / Exact Map of / Darbieshire / Corrected and Amended with / Additions By P. Lea.

10. The / County Palatine / of Chester by C.S. / Corrected and Amended / with many Additions / by P. Lea.

11. Stafford / Discribed by C.S. / Corrected and Amended / with many Additions / by P. Lea.

12. Shropshire / accurately drawn / and sett forth / by C.S. Corrected, with some / Additions by P. Lea.

13. The County of / Hereford / resurveyed / and / enlarged. / An^o: 1665.

14. Worcestershire / and Citty / exactly described by C.S. / newly / augmented / by Phil: Lea.

15. Warwick / and / Leicester / Shires / Described by C.S. / Corrected & amended wth / many additions by P. Lea.

16. The County of / Northampton, / togeather wth ye three small / Counties of / Bedford Hvntingdon / & Rvtland, / exactly drawn by one Scale / by C.S. Corrected & Amended / with Many Additions by P. Lea.

17. The / Covnty of / Norfolke / exactly plotted / By C.S. Corrected and / amended By P. Lea.

18. Svffolke / described / By C.S. Corrected and / Amended by P. Lea.

19.* Essex / Actually surveyed / with the several Roads / from London etc. / Exactly measured at three / mile in an Inch. / Sold by Phil: Lea at ye Atlas / and Hercules in Cheapside / London. *In the bottom right-hand corner is the dedication:* To the Right Hono:^{ble} / Arthur Earl of Essex / Viscount Maldon Baron of Hadham / Lord Leivetenant of the County of Hertford etc. and one of His Maje:^{ties} / Most Hono.^{ble} Privy Councel / This Map actually Survey'd is Humbly presented and Dedicated / by yor Hono.^{rs} most obedient serv.^{t} / William Morgan / His Maties Cosmographer / & M^r. of ye Revells / in Ireland. (*The dedication shows that this particular map was engraved not later than 1683*).

20. Hartford / Shire / by C.S. Corrected / and Amended / with many Additions / by P. Lea.

21.* An Actuall Survey of / Midlesex, / Sold by Phil:·Lea at the Atlas and / Hercules in Cheap: side / London. / Walter Binneman sculp. A scale of English miles, 5 (=82 millimetres). *In the top left-hand corner is the dedication:* To the Ho.^{ble} / S^r. Thomas Wolstenholm / of Minsenden in com. Midd. Bar^t. / This new Map of / Midlesex / actually survey'd and deliniated / is humbly Presented & dedicated / By John Ogilby Esq. / his Maties / Cosmographer.

22. Oxford, / Buckingham / & / Bark-shire. / By C:S: Corrected and / Amended with many / Additions by P: Lea.

23. Glocester-Shire / Discribed by C.S. / Corrected and Amended / with many Additions by P. Lea.

24. Monmovthshire / heretofore part / of Wales, / But now added / to other English / Shires in ye Con: stant Circuite / of the Iudges by C S / Corrected & Amended / By P Lea.

25. Cornwal / with y^e severall hundre / ds, truly described / by C.S. Corrected and / Amended by P. Lea.

26.* Devon-shire / Described by C. Saxon. / Corrected, Amended, and / many Additions by P: Lea. / Francis Lamb Sculp. A scale of miles, 10 (= 67 millimetres).

27. The County of / Somerset / exactly plotted. / Published An° 1689.

28. Wiltshire wth Salisbvry Citty / & Stone heng described An°. 1689.

29. A true description / of / Dorsetshire / By, C.S. Corrected / and Amended With / many Additions / By P. Lea.

30. Hampshire / by / C: Saxton. Corected & many / Aditions by P: Lea.

31. Sussex, Surry, / and Kent, by C.S. / Corrected & Amended with / many Additions / by Phil: Lea.

32. Mona Insula alias Anglesey / and Caernarvan Shire / Discribed by C.S. Corrected and / Amended with many Additions / By Phil: Lea.

33. Denbigh / And / Flint Sh = / Discribed by C.S. / Corrected and Amended / With many Additions / By P. Lea.

34. Merioneth / And / Montgomery / Discribed by C.S. / Corrected and Amended / By P. Lea.

35. Radnor, Breknoke, / Cardigan And / Carmarthen, / Shires Discribed / By CS. Corrected / and Amended with / many Additions / by P. Lea.

36. Pembrock / Shire / Discribed by C.S. / Corrected and Amended / with many Additions by P. Lea.

37. Glamorgā / Shire Discribed by / CS, Corrected and / Amended by P. Lea.

Maps missing from the Royal Geographical Society's copy
I.* A Map of the / Isle of Wight / Portsea Halinge, also / The islands of Iarsey and Garnsey / Which are a Part of / Hampshire. / Made & sold by Philip Lea.

II.* Cambridge = Shire / and / The Great Levell of y^e Fenns, / extending into the Adjacent Shires, / according to Surveys as is now drained, / at y^e Charges of y^e R^t. Ho.^ble W. Earl of Bedford, / and y^e other Proprietors by S^r. Jonas Moore, etc. / Made and sold by P: Lea, at y^e Atlas and / Hercules in Cheap:side. London.

TABLE 8

Maps in the Saxton Atlas by Lea, c.1693

Sequence as in the British Library's copy (Map Room: c.7c.4). Maps which do not derive from Saxton's plates are indicated by an asterisk. Maps belonging to the edition, but missing from the British Library's copy, are appended at the foot of the table in no particular sequence.

1. Anglia / The Kingdome of / England / and Principality of / Wales / exactly Described / By Chr: Saxton. / P. Lea excudit.

2.* Cambridge = Shire / and / The Great Levell of y^e Fenns, / extending into the Adjacent Shires, / according to Surveys as is now drained, / at y^e Charges of y^e R^t. Ho.^ble W. Earl of Bedford, / and y^e other Proprietors by S^r. Jonas Moore, etc. / Made and sold by P: Lea, at y^e Atlas and / Hercules in Cheap:side. London.

3. The / County Palatine / of Chester by C.S. / Corrected and Amended / with many Additions / by P. Lea.

4. Cornwall / Described by C: Saxton / Corrected & many Additions as / the Roads etc. by P: Lea.

5. Comberland / and / Westmorland / Exactly described / By CS Corrected & Amended / with many Additions by P Lea.

6. DerbyShire / Described by C: Saxton Corrected & / Amended with many Additions as / Roads etc. by P: Lea.

7.* Devon=Shire / Described by C: Saxton / Corrected, Amended, and / many Additions by P: Lea. / Francis Lamb Sculp. A scale of miles, 10 (=67 millimetres).

8. Dorsetshire, / Described by C: Saxton. / Corrected and Amended / with many Additions / as Roads etc. by P: Lea.

9. The County Palatine and Bishoprick of / Durham / Described by C: Saxton Corrected and / Amended with Additions by P: Lea.

10. A Map Of / Essex newly revised / & Amended by P Lea.

11.* Essex / Actually Surveyed / with the several Roads / from London etc. / Exactly measured at three / mile in an Inch. / Sold By Phil: Lea at ye Atlas / and Hercules in Cheapside / London. *In the bottom right-hand corner is the dedication:* To the Right Hono:^{ble} / Arthur Earl of Essex / Viscount Maldon Baron of Hadham / Lord Leivetenant of the County of / Hertford etc. and one of His Maja:^{ties} / Most Hono.^{ble} Privy Councel / This Map actually Survey'd is Humbly presented and Dedicated / by yor Hono.^{rs} most obedient serv.^{t} / William Morgan / His Maties Cosmographer / & M.^{r} of ye Revells / in Ireland . . . F. Lamb sculp.

12. Glocester-Shire / Described by C.S. / Corrected and Amended / with many Additions by P. Lea.

13. Hampshire / by / C: Saxton Corrected & many / Aditions by P: Lea.

14. The County of / Hereford. / resurveyed / & / enlarged / An^o: 1665.

15.* To his Grace / the most Reverend Father in God / Iohn Tillotson by divine Providence / Lord Archbishop of / Canterbury / Primate of all England and / Metropolitan, & / one of his Majesties most honourable / Privy Council, etc. / This new Map of / Kent / is most humbly Presented and Dedicated, by / P. Lea.

16. The County Palatine of / Lancaster / Described by C: Saxton Corrected and / Amended with Many Additions of / Roads etc. by P: Lea. *The copy in the Manchester Public Library has the title:* The / County Pallatine / of / Lancaster.

17. Lincolne Shire / And / Nottinghame Shire / By. C Saxton.

18.* Midlesex / Actually surveyd and / Deliniated / By John Seller Hydrographer to y^e King / Cum Pre=vilegio Regis. / Sold by Phil: Lea at y^e Atlas & Hercules / in Cheapside.

19.* An Actuall Survey of / Midlesex, / Sold By Phil: Lea at the Atlas and / Hercules in Cheap=side / London. / Walter Binneman sculp. A scale of English miles, 5 (=82 millimetres). *In the top left-hand corner is the dedication:* To the Ho.^{ble} / S^r. Thomas Wolstenholm / of Minsenden in com. Midd. Bar^t. / This new Map of / Midlesex / actually survey'd and deliniated / is humbly Presented & dedicated / By John Ogilby Esq. / his Maties / Cosmographer.

20. Monmovth shire / heretofore part / of Wales, / But now added / to other English / Shires in ye Con: / stant Circuite / of the Iudges by CS / Corrected & Amended / By P Lea.

21. Norfolk / Described by C: Saxton / Corrected and Amended with / many Additions of Roads etc. / by P: Lea.

22. The County of / Northampton / togeather wth ye three small / Counties of / Bedford Hvntingdon / & Rvtland, / exactly drawn by one Scale / by CS, Corrected & Amended / with many Additions By P Lea.

23. Northumberland / Described by C: Saxton / Corrected and Amended / By Phill: Lea.

24. Oxford / Buckingham / & / Bark-Shire. / By C:S: Corrected and / Amended with many / Additions by P: Lea.

25.* A / Map of / Oxford / Shire / With the Roads Sold by Philip Lea at the Atlas & Hercules in Cheapside / London.

26. Shropshire / accurately drawn / and sett forth / by CS Corrected, with some, / Additions by P Lea.

27. Sommersetshire / Described by C: Saxton / Corrected and Amended with / many Additions as Roads etc. / by P. Lea.

28. Stafford. / Discribed by C.S. / Corrected and Amended / with many Additions. / By P. Lea. (*Saxton's Latin title remains in a copy in the University Library, Cambridge.*)

29. Suffolk / Described by C: Saxton / Corrected & Amended with / many Additions as Roads etc. / by P. Lea. (*In the copy in the Manchester Public Library, the title is* 'Svffolke / described').

30.* Surrey / Actually Survey'd and / Delineated / By John Seller / Hydrographer to the King / many Additions by P. Lea / Cum Privilegio Regis ... Iohn Oliver & Richard Palmer Sculpsit ... Sold by Philip Lea at ye Atlas and Hercules in Cheap Side.

31. Sussex, Surry / and Kent, by C.S. / Corrected & Amended with / many Additions / by Phil: Lea.

32. Warwick / And / Leicester / Shires / Described by C.S. / Corrected & amended w:th / many additions by P. Lea.

33. Wiltshire, wth Salisbvry Citty / & Stone heng described An:º 1689.

34. Worcester Shire / Described by C: Saxton Corrected / and Amended with many Additions / as Roads etc. by P: Lea.

35. York-Shire / Described by Ch: Saxton. / Many additions, and Corrections as / y^e Roads, wapentakes etc, by P. Lea.

36. Mona Insula alias Anglesey / And Caernarvan Shire / Discribed by CS Corrected and / Amended with many Additions / By Phil: Lea.

37. Denbigh / And / Flint Sh= / Discribed by C.S. / Corrected & Amended / With many Additions / By P. Lea.

38. Merioneth / And / Montgomery / Discribed by C.S. / Corrected and Amended / By P. Lea.

39. Pembrock / Shire / Discribed by CS / Corrected and Amended / with many Additions by P Lea.

40. Radnor, Breknoke / Cardigan And / Carmarthen, / Shires Discribed / By CS, Corrected / and Amended with / many Additions / by P. Lea.

Maps missing from the British Library's copy

I.* A Map of the / Isle of Wight / Portsea Halinge, also / The islands of Iarscy and Garnscy / Which are a Part of / Hampshire. / Made & sold by Philip Lea. *The page includes maps of* 'Holy Island', 'Farne Island', 'The isle of / Man', 'The Islands / of Scilly / Surveyed by Cap. / Collins', *an untitled map of the Isle of Wight,* 'Garnsey', *and* 'Iarsey'.

II.* A Mapp / Containing the / Townes Villages / Gentlemens Houses / Roads Rivers Woods / and other Remarks / for 20 Miles Round / London / ... Sold by Phil. Lea at ye Atlas & Hercules in Cheapside London.

III. Hartford / Shire / by C:S: Corrected / and Amended / with many Additions / by P: Lea.

IV.* Hertford / Shire / Actually Survey'd and / Delineated / By John Seller / Hydrographer to the King / Cum Privilegio Regis. / ... To His Excellency Arthur Earle of Essex Viccount Maldon ... This Map is humbly Dedicated by John Seller / Hydrographer to the King.

V. Glamorgā / Shire Discribed by / CS, Corrected and / Amended by P. Lea.

VI.* Scotia Regnum cum insulis adjacentibus. Robertus Gordonius a Straloch descripsit.

VII.* A New Map of / Ireland / According to Sr W. Petty (but supplied wth Many Additions / which are not in his Survey, nor in any other map) Divided into / Provinces, Counties & Barronies, where in are / Distinguished not only the Bishopricks & Borough^s but also / all the Bogs, Passes, Bridges, etc; y^t are in S^r; W: P. 32 County Maps. / To their Most Excellent Majesties / William and Mary / King and Queen / of Great Britain, France and Ireland etc. / And to their Most Hono^{ble} Privy Council / This Map is humbly Dedicated by Your Majesties / Most Dutiful & Loyal Subjects. P. Lea and H. Moll. *The map is divided into two sheets, viz:* 'The North Part of / Ireland ... H. Moll. Fecit', *and* 'The South Part of / Ireland'. (*The map is undated.*)

VIII.* *A page with five maps, viz:* 'Dover and the Downes'; 'London'; 'Harwich'; 'The East Part of the River Thames etc'; 'The West Part of the River Thames'.

IX.* Buckingamia / Comitatvs / vulgo / Buckingham / Shire / By Iohn Seller ... Corrected & many Additions / By P Lea: ... / Sold by Philip Lea at the Atlas & / Hercules in Cheapside London.

Chapter 6

Rewards and recognition

In Elizabethan England the establishment functioned by a system of dispensation of patronage, which emanated from the Queen to her courtiers and principal ministers of state, and through them percolated down to the lesser gentry. It was a system which worked well. By this means the Queen ensured loyalty from her politically conscious subjects, and many of the rewards bestowed cost her nothing in financial terms.

Patronage from the Queen took many forms: grants of honours, administrative offices, leases of crown lands on favourable terms, and grants of monopolistic patents were typical ways in which she rewarded her favoured subjects. Christopher Saxton benefited in all these ways and it seems that the Queen followed his career with close interest.

It has generally been understood that Thomas Seckford paid Saxton for his work and met all his expenses. The proofs of Saxton's maps were given to Lord Burghley as they were completed from 1574 onwards, and these were kept by Burghley and later bound together, with other maps, for his personal use. The Royal Arms were engraved upon every map, excepting that of Norfolk. It is apparent that Burghley kept the Queen informed of progress and she in turn showed her approval and encouragement of the project by requiring the Royal Arms to be engraved upon the maps.

According to Dr Favour's 'epitaph' described on pages 6-7, Saxton received 'letters' from the Queen on July 28th, 15 Elizabeth (1573), which cannot now be traced. Presumably it was authority for Saxton to carry out his survey, possibly requiring local officials to offer any assistance necessary to complete the work. By that date, however, Saxton must have already undertaken much of the groundwork and the 'letters' must have been given as encouragement and a sign of approval.

There are two surviving records of official instructions for assistance to be given Saxton in his survey. The first is dated March 11th, 1575, when his survey was well under way, and required that Saxton, 'servnt to Mr Sackeford' be assisted wherever he went (Appendix 15). The second, dated July 10th, 1576, took the form of an open letter to Justices of the Peace, Mayors and others in Wales requiring them to aid and assist Saxton during his survey of that country, 'to see him conducted unto any towre Castle highe place or hill to view that countrey'. It was also requested that he be accompanied and assisted by two or three honest men that knew the country. On his departure from a town Saxton was to be conducted by a horseman who spoke both Welsh and English (Appendix 8).

The letter is the only direct reference to Saxton's procedure and method employed in surveying. He appears to have travelled alone employing local assistance.

It is clear that in the early days the credit for the maps was given to Thomas Seckford. For instance Raphael Holinshed in his *Chronicles*, written in 1577–86, says: '... vnderstanding the great charges and notable enterprise of that worthie Gentleman maister Thomas Sackford, in procuring the Charts of the seuerall prouinces of this realme to be set foorth, we are in hope that in time he will delineate this whole land so perfectlie, as shall be comparable or beyond anie delineation heretofore made of anie other region; and thefore leaue that to his well deserued praise.' William Harrison, in dedicating his description of Scotland to Thomas Seckford in the same *Chronicles*, writes: 'Hauing by your singular curtesie receiued great help in my description of the riuers & streames of Britaine, and by conference of my trauell with the platforms of those few shires of England which are by your infinite charges alredie finished (as the rest shall be in time by Gods helpe, for the inestimable benefit of such as inhabit this Iland.)'

Although Seckford received royal patronage, none is specifically noted as having been due to his initiative in financing the survey.

Christopher Saxton himself received his first mark of appreciation on March 11th, 1573, when the Queen 'in consideration that Christofer Saxton for certain good causes grand charges and expenses lately had and sustained in the survey of divers parts of England' granted him an estate in Suffolk previously held by Lady Anne of Cleves, divorced wife of Henry VIII. The lands were in Stafford, Benall, Fernaham, Great Glenham, Little Glenham, Swefflinge and Rendham, and were known as Grigston lands or Grigston manor, bringing in an annual rent of £9 18s 3d. There was also a field in Stratford that brought in 12d. With the manor went the privilege of holding courts leet and views of frankpledge with the profits to be gained therefrom. Exclusions, such as wards, marriages, minerals and advowsons were specified. The grant was for twenty-one years, Saxton to pay £10 5s 11d annually to the crown. In comparing the basic income from rent and the sum payable to the crown it is obvious that there must have been other income available from the estate; a certain amount would come from the courts held and some from the mills and admittance fines (Appendix 7).

The lands were scattered through five parishes, but all in Plomesgate Hundred, to the west and south-west of Saxmundham in Suffolk, roughly ten miles from Woodbridge, home of Thomas Seckford.

Grigston, or the manor of Griston as it became known, came to the crown on the dissolution of the monastic foundations and was granted by Henry VIII to Anne of Cleves for life. On her death in 1557 it reverted to the crown. The subsequent history of the manor is confused, but it is likely that it was possessed by Thomas Glenham until his death in 1571 and then, on the expiration of Saxton's grant, was granted to Richard Firth and Edward Hawtayne.[1]

On January 19th, 1574, the Queen made Saxton a further grant, this time 'in consideration of the services of Christofer Saxton in and about the survey and description of all and singular the counties of England . . .' It was the office of bailiff and collector and receiver of rents and profits of the manors, property and land in the city of London and Middlesex which had previously belonged to the Priory or Hospital of St John of Jerusalem. The grant was not immediate, but on the death or forfeiture of 'Constant Benet' who then held the office. Thereafter it was Saxton's for life. An annual fee of £10 was payable to the Exchequer from his profits (Appendix 16). Constance Bennet, a Greek, died in 1577 so Saxton had not long to wait to assume his mantle of office.

Again we cannot guess at the income derived from this office. On the suppression of the Grand Priory in St John's Gate, Clerkenwell, it was valued in 1540, according to Stow 'to dispend in lands three thousand three hundred and eighty five pounds nineteen shillings and eightpence yearly'.[2] Clearly the priory had owned a considerable amount of property; records survive of properties owned in the City and Middlesex at the dissolution, but it is not known how many had been granted or sold by the crown in the period up to 1575, although it is likely that a considerable proportion were still in hand at the date of Saxton's grant. The position is complicated after the dissolution because of reversals of policy by Mary Tudor and then Elizabeth concerning the religious foundation. In the 1570s the Hospitallers probably owned property in over fifty places in London and

67

Middlesex. Evidently Saxton held a profitable office; in all likelihood he would employ a deputy to collect the rents on his behalf.

On July 19th, 1577, the Queen granted Letters Patent to Saxton giving him the sole right and licence to print and publish his maps, both those already completed and others he should print, for a period of ten years. It is a lengthy legal document addressed to printers, booksellers and others stating that Christofer Saxton servant of Thomas Sekeford Master of Requestes, had already, at great cost to his master, travelled throughout England and drawn and set out true and pleasant maps of the counties which had been engraved. For his encouragement to proceed the Queen granted to Saxton the privilege and licence to both print and sell any of his maps for a term of ten years. At the same time all others, printers, booksellers or foreigners were forbidden to print the maps: if they did so it was 'uppon peyne of oure hieghe indignacon and displeasure', and the offender would incur a fine of £10, and forefeit all such maps illegally held or printed (Appendix 9).

This licence was valuable at a time when plagiarism was the order of the day. It would enable Saxton to profit considerably from his undertaking, as was the intention. At this date the maps were sold individually as available, the atlas bound as a whole not being published for another two years.

It may be wondered why the licence was granted in 1577 instead of 1579, but this was because Holinshed's *Chronicles* was published in 1577 and it was anticipated that there would be a demand for Saxton's maps to accompany the work. On the other hand it did lessen the period to eight years in which Saxton had sole right to print and sell following publication of the complete atlas; thereafter any printer or bookseller could deal in the maps or use as desired. The information contained in Saxton's maps was taken by subsequent map-makers, generally without acknowledgement, for many generations to come.

Nothing is known of Saxton's method of sale, or whether he used a book or printseller as intermediary. An atlas was purchased direct from Saxton for 76s 6d for the chamber of the Duchy of Lancaster at Westminster. The volume, described as 'les mappes', was 'for the better instruction of the Queen's officers in such causes as in that book were mentioned'.[3] Fordham records that in 1736 the atlas was selling at 15s to 20s. Today, four hundred years after publication, it realises about £30,000.

Having thus received grants of office, land and a patent prior to the actual publication of his atlas, Saxton was further honoured by the Queen in the year in which his atlas appeared. He was granted amorial bearings. He had 'arrived'. The record of this grant is contained in two manuscripts at the Bodleian Library, Oxford, one being a copy of the other. The earliest was written by the herald Robert Glover, who lived 1544–88 and who must therefore have made his copy from the original at a contemporary date (Appendix 17). The second is a copy of Glover's entry made by Elias Ashmole in the seventeenth century.[4]

The grant, dated July 1st, 1579, was confirmed by William Flower, Norroy King of Arms of the North Parts of England. It states that he was required by Christopher Saxton to describe for him the arms of his ancestors that he may lawfully use and bear. In addition, in respect of Saxton's worthiness, 'who by speciall direction and commandment from the Queens Matie hath endevored to mak a perfect geographicall description of all the severall shires & countis wthin this realme and accordingly finished the same to his everlastinge prayse'. He was to have a personal crest surmounting the arms of the forearm of a man holding a pair of open compasses. A sketch of the arms and crest accompanies the text.

The original grant at the College of Arms cannot be traced, nor any record of Christopher Saxton's ancestral pedigree as was usually required in claiming ancient family arms, although in special circumstances such as this that requirement might be overlooked.

Arms of Christopher Saxton.

However, Sir George Fordham remedied at least part of this by having Saxton's arms redrawn and certified by Arthur Cochrane, the Norroy King of Arms, in January 1927.

Although Saxton's immediate family has only been traced back two generations, the arms he took were ancient arms of the Saxtons and most probably those of his own ancestors through a lesser line of the family. There are in fact pedigrees of two families bearing these arms, but no connection has yet been discovered linking either of them with the Saxtons of Dunningley (Appendix 18).

Exactly one year later, on July 1st, 1580, the Queen made Saxton a further grant of land, this time for a period of sixty years. It was a piece of waste ground in St John Street in the parish of St Sepulchre without Newgate, in the City of London. It lay between St John Street on the east and Turnill Street on the west, to the north of West Smithfield in the manor of Clerkenwell. The exact location is impossible to pinpoint, nor can the exact size of the plot of land be ascertained for, although the length of each side is specified, the measurements are in virgates and these were variable in size in different parts of the country, and were normally units of acreage. Saxton was to pay 3s 4d annually for the land and had permission to build one or more houses thereon (Appendix 19).

Some eleven days later the Privy Council issued an open letter to all Justices of the Peace, Mayors and Sheriffs stating that the Queen had granted the above land to Saxton with the intention of him building 'certen convenient buildings there, as, namelie, a

Sessions Howse for the Justices in Middlesex to kepe Sessions of Enquirie for that countie'. There had been a move in recent years to curtail building within London in an effort to reduce overcrowding, and the Privy Councillors' letter continues to say that despite the late proclamation, Saxton should not be in any way impeached, but be allowed to proceed according to his grant. Their Lordships, in her Majesty's name, commanded those 'to whom it maie appertaine quietly to suffer the said Christofer and his workmen to fynishe the same, notwithstanding this Proclamacion' (Appendix 20).

There is no record of a Sessions House constructed by Saxton under this lease; the Middlesex Justices met at that time in the Castle Inn in St John Street. The first Sessions House was built in 1609 by Sir Baptist Hicks in that street. What, or even if, Saxton built on his plot is not known, but according to Stow writing a few years later, St John's Street was 'on both the sides replenished with buildings up to Clerkenwell'.

A further favour received by Saxton from the Queen was a lease of the Rectory and Church of Scalby in Yorkshire for a term of twenty-one years. The lease, contained in Letters Patent, was to run from 25th March, 1584 (although the enrolment of the Letters Patent was on 27th June of that year), but no rent was payable until 25th March, 1588, when Saxton was due to get formal posssesion, unless that happened in the meantime. The Queen, prompted by Lord Burghley, Treasurer of England, and Sir Walter Mildmaye, Chancellor of the court of Exchequer, granted to Christopher Saxton, 'gentleman', the Rectory and Church of 'Scawbye' in the county of York, then or formerly in the tenure or occupation of William Clifton and John Singleton and prior to that belonging to Bridlington Priory. The lease included buildings, land and tithes, both great and small, with stated exceptions such as woods, mines, quarries and the advowson of the vicarage. There were certain obligations such as the maintenance of the chancel of 'Scawbie' church together with other buildings, hedges, seawalls, etc., for which Saxton was allowed to take timber. For all this Saxton was to pay £10 10s 0d annually. There must have been quite a considerable income to be gained, as the sum Saxton had to pay was in excess of that he had to submit for any of his previous grants (Appendix 21).

Scalby, as it is now called, is two miles north-west of Scarborough in the North Riding of Yorkshire. The church is in the patronage of the Dean and Chapter of Norwich. It is rather curious that in the archives of the Dean and Chapter of Norwich there is no record of this lease to Saxton and in fact there are records of the rectory and advowson of Scalby being leased to others at the same period.[5]

Another activity which engaged Saxton's attention, although of short duration, was his appointment as bailiff of the Duchy of Lancaster for West Pontefract and the Wapentakes of Agbrigg and Morley in the West Riding of Yorkshire in 1589. There are several papers extant concerning Saxton's appointment, which was to replace John Foxcroft appointed in 1580, but who had been dismissed. On November 3rd, 1589, a draft patent was issued stating that, by the agreement of the Council of the Duchy, they granted the office 'to our beloved Christopher Saxton, gent'. He was to occupy the office during the Queen's pleasure, which meant indefinitely, with the customary wages, fees and profits. The revenue was to be paid to the Particular Receiver twice a year and Saxton, or his deputy, had to attend the assizes and sessions for the county, as was customary for the bailiff, and render annual accounts to the Auditor of the Duchy. The document was signed by Sir Francis Walsingham, Chancellor of the Duchy and Joint Secretary of State with Lord Burghley, and John Brograve, Attorney General, on behalf of William Gerrard, Clerk of the Council. The format was standard, but does throw a little light on the duties of a bailiff (Appendix 22).

Accompanying the above is a sheet of accounts, known as the particular, rendered by

John Foxcroft. At the bottom of the page, dated October 18th, 1589 (before the Letters Patent were granted), is written the Chancellor's authority to the Clerk of the Council to make a fair copy of the patent and to take a bond from Saxton as surety.[6]

The Chancellor's Warrant to Gerrard, again signed by Walsingham, states that he is required to take a bond from Saxton 'in the some of fiftie pounds' that he shall truly discharge himself, and that two others be bound with him. Three men are nominated as acceptable, Robert Mawe of Lincoln's Inn, Peter Page and Thomas Hall (Appendix 23). In fact only the first two were called upon and a further entry, on November 3rd, 1589, records receipt of £500 'for the answeringe his office' when Saxton is again described as of 'Dunninglawe, gent'. There is a discrepancy, but it is believed that the latter was in fact an error in entry and that £50 was the true figure.

Robert Mawe was a Suffolk man, a barrister, admitted at Lincoln's Inn in January 1576–7 from Furnivall's Inn. Thomas Hall, from Gloucester, followed the same profession, being admitted to Lincoln's Inn in November 1579 from Thavies Inn. Peter Page on the other hand was Saxton's brother-in-law, being married to Margaret Saxton.

Saxton held the office of bailiff for one year only, surrendering his letters patent for cancellation prior to the appointment of Thomas Clayton on November 28th, 1590.[7] During Saxton's period of service Gilbert Talbot was Steward of the Honour of Pontefract and Constable of Pontefract Castle, and Sir John Savile of Howley and Edward Carey were joint Stewards of the Manor of Wakefield, which areas basically covered the territory designated to Saxton as bailiff.

According to Somerville, bailiffs ranked among the officers in the hierarchy of appointees of the Duchy Court; the Chancellor generally made the choice, but these were sanctioned by the Queen. An appointment could be the equivalent of the grant of an annuity or pension in respect of past or future services, and from an accounting officer there would be expected honesty, regular rendering of account and payment of money.[8]

That Saxton's office involved him in considerable financial dealings is clearly demonstrated by the comparatively large surety required. The amount of the security was not basically related to the value of the office to the office-holder, but to the amount of money involved in his transactions, because the Duchy wished to be able to recoup its losses if the officer defaulted. The security money was not usually paid, but was the sum the Duchy could claim, or sue for, on default.

Saxton's status within his own parish of Woodkirk is demonstrated by his nomination as a trustee of some almshouses at Westerton in Woodkirk, established by Robert Greenwood. The trust deed, dated August 28th, 1593, states that Robert Grenewodde of Westerton, gentleman, conveyed to James Greenwood his son, Henry Batt of Topcliffe, Joseph Peck, Christopher Saxton and Alveraye Linleye of Dunninglawe, Nicholas Linleye of Baghill, George Farebank of Tinglawe, Thomas Wholeye, Edward Crowder of Wether Green and Edward Purston of Westerton, yeoman, a parcel of land with a newly built house thereon, with adjoining garden in Westerton. This was the almshouse in which three nominated poor women were to reside for life. A second house in Westerton was let at 20s rent to provide an annuity.[9]

The Charity Commissioner's Return of 1899 details the later history of Greenwood's Hospital as it came to be known. From 1892, when the last occupant died, no one could be induced to live there, the building being then very dilapidated, although it was thought that the original house had been rebuilt or much renovated in the early nineteenth century.

Robert Greenwood of Westerton, a barrister, was appointed Under Steward of the Manor of Wakefield in 1587. Henry Batt, another of the trustees, was a Deputy Steward of the Manor of Wakefield and was responsible, in 1569, for handing over the Court Rolls for safe custody at Sandal Castle. George Farebank was an attorney and the Linleys were a well respected local yeoman family. Trustees were obviously men of standing.

Christopher Saxton as a relatively wealthy man had a finger in another pie, entirely different to any of his other activities, and probably purely a financial venture. He held a half share of Adwalton Fair, with the right to collect tolls and hold a Court of Piepowder.[10] Adwalton, in the parish of Birstall and almost four miles north-west of Dunningley, had had a fair from time immemorial. It had become the haunt of horse thieves and in February 1576, with the object of controlling the trade, the Queen ordered Letters Patent to be issued under the Duchy seal to James Brooke, owner of the manor of Drighlington and licencee of The White Hart Inn at Adwalton, adjacent to which the fair was held. James Brooke was granted the right to hold a market every second Thursday, and two annual fairs. He died shortly after however, and his son John Brooke purchased a new grant on February 1st, 1577, on the same terms.[11] Thirteen days later Brooke sold a moiety, or half part, of his rights in the markets and fairs to Christopher Saxton, 'gent'. This share was held by Saxton for the rest of his life and was eventually sold by his son Robert Saxton, in 1616, to William Brooke, John's son, and Thomas Hirste.

Included in the grant was the right to hold 'a Court of Piepoudre to be holden there at the time of the said fairs, feasts, and markets with all liberties and free customs to the same Court belonging and with tollage, stallage, pannage, fines, americaments and all other profits, commodities and emoluments whatsoever ..'. The grantees were to collect '... for every lamb bought or sold in Adwalton one penny, and for every live sheep there bought or sold one penny, and other dues tolls and imports of all manner of merchandizes and chattels whatsoever sold or bought there ...'. For this right Brooke was to pay 26s 8d annually to the Court of the Exchequer.

It must have been a profitable investment, for it is recorded in 1631 that buyers came from Derbyshire, Lancashire, Cumberland and Westmorland as well as all parts of Yorkshire. Presumably Saxton would employ a deputy as he spent so much time away. A strong link developed between the Saxtons and Brookes: Thomas Saxton, Christopher's brother, married Alice Brooke and named John Brooke as supervisor of his Will.

In 1629 there was trouble in Adwalton and resultant litigation throws considerable light on the markets and fairs there. The inhabitants of Bradford had been trying to divert trade from Adwalton to Bradford, employing all manner of subterfuge to do so before the Brookes finally resorted to law. Interrogatories were drawn up which included the question 'Do you know or are you a witness to or have you seen any grants, assignments or writings whereby John Brooke of Adwalton, did grant a moeity of the fairs at Adwalton to one Christopher Saxton, or whereby the said Christopher Saxton and one Robert Saxton, his son, did grant or assign the said moiety to the complainant William Brooke of Drighlington ...'. Another read: 'Do you know that the said John Brooke of Adwalton, ever since the grant by Letters Patent from Queen Elizabeth made to him in or about the nineteenth year of her reign, and the same Christopher Saxton and the said complainants have held and enjoyed the fortnight and other fairs at Adwalton, and taken the tolls and profits thereof and for how many years is it to your best remembrance that they or some of them have held the same.' Depositions of witnesses were taken on September 28th, 1629, when an octagenarian, William Hall of Drighlington, said that he had been witness to deeds of sale of the moiety, or half part, of the markets and fairs made by Christopher Saxton, and Robert his son, to William Brook of Drighlington, and his heirs. The indenture of sale was dated about January 10th, 1616. Since about the nineteenth year of Queen Elizabeth, according to Hall, the Saxtons, John Brooke of Adwalton and the complainants had held the markets and fairs of Adwalton and had taken the profits thereof to their own use.

Another witness, Raphe Beeston of Tong, said that he had seen an indenture dated February 13th, 1577, whereby John Brooke conveyed a moiety of the profits of the fairs and markets of Adwalton to Christopher Saxton, gent. A toll keeper stated that he considered the toll of Adwalton fair was worse by £10 as a result of the Bradford men's actions. There was evidence that the latter were willing to spend £1,000 to get the fair diverted to Bradford. If that were the case, it must certainly have been a source of considerable revenue to the Brookes and Saxtons in their day.

The deposition of William Hall is interesting in another respect; he says that Christopher Saxton was a party to the sale of his share in January 1616. All other evidence indicates that Saxton was dead by then, and the feet of fines records only Robert Saxton and his wife Ann involved in the sale, so not much credence should be placed on the doubtful memory of an aged witness.

Chapter 7

Christopher Saxton: the estate surveyor

Following the publication of his county atlas and wall-map Saxton must have wondered, what next? He had become something of a celebrity as his maps circulated and the novelty of 'seeing' the country engraved on paper reached many quarters for the first time.

The redistribution of monastic lands and subsequent acquisition of new estates by the gentry, together with the enclosure of the common lands in the sixteenth century, created a demand for large scale surveys of lands and boundaries of a more sophisticated nature than the previously adequate written 'extents'. It was as a surveyor in this field that Saxton carved his niche in later life.

From previous lack of evidence to the contrary it was presumed that Saxton sank into relative obscurity, retiring to his home at Dunningley, occasionally being employed to prepare a map or survey of some tract of land. This was an understandable assumption, for when Fordham wrote his life of Saxton in 1927 only six manuscript maps, including evidence of one lost, and one written survey were known.[1] Over the years many more have come to light and evidence of others, now lost, has been discovered. In 1972 twenty-two maps and one written survey were listed.[2] Now, twenty-five maps or authentic copies and fourteen written surveys are known, and there is evidence of a further eleven having been prepared. The very nature of maps and surveys rendered them vulnerable to destruction, particularly as the content rapidly became outdated or superseded. Those that survive therefore may well be only a fraction of the number originally made and Saxton was probably fully employed as a surveyor.

Whether Saxton worked alone, had a permanent assistant, or engaged help locally for each task is not known, except in a few instances when either another surveyor is named, as at Whitchurch, or instructions were given for a layman to assist him, such as the clerk of St Thomas' Hospital when he surveyed their estates. The fact that many of the surveys are not in Saxton's hand suggests that others were involved, but as the script varies it was obviously not a single man throughout the whole period. Robert Saxton assisted his father when of age and eventually worked independently, although never undertaking any project of great significance.

In only one instance, Thornhill, do both a map and written survey survive. It is probable that both were made in many cases, particularly where the survey was of an estate, as sufficient information is rarely given on the maps alone to render them comprehensive. In Robert Saxton's case both maps and surveys survive of Manningham, Tanshelf and Gomersal.

Techniques of large scale surveying had become well established by the last quarter of the sixteenth century (see pages 42-45), and a comparison of Saxton's manuscript maps with modern large scale maps reveals a surprising accuracy, even more surprising when it is observed how inaccurate were the numerical additions on his surveys. Christopher Saxton invariably used 'daworks', as he spelt it, as a unit of land measure. There are four perches in one daywork, ten dayworks in one rood, and four roods in one acre. This was a specific measurement rarely used outside Kent and Essex, recorded in use in the latter as early as

1200, but which had according to Fisher, died out there by the time of Queen Elizabeth.[3] Only one other contemporary surveyor working in Yorkshire used the four perch daywork, and he was a Kentish man. Only twice did Robert Saxton use dayworks, in surveys of Sandal and Manningham, otherwise preferring the more universal measure. The result was the same.

Although the statute measure of a perch was five and a half yards, or sixteen and a half feet, variable lengths were still employed and acceptable and Saxton was known to have used a perch of twenty-one feet when he surveyed Methley. Instances of as little as nine feet or as much as twenty-six feet have been recorded,[4] and even in the seventeenth century variable perches were still fairly commonly used, ranging from the perch of twelve feet, called tenant or court measure, to a perch of eighteen, twenty, or even twenty-four feet, called the woodland measure (Appendix 24). Though Saxton invariably gives the scale of a map (excepting on that of Ribston which is a copy), it is generally a scale of perches, without indicating which length of perch is utilised (except on the maps of Ribston and Hunsingore where it is the statute measure, and it is questionable whether this explanation was original or the idea of the copyist [see page 110]). Thus unless a direct comparison between any of Saxton's maps and the actual known size is made, it cannot be ascertained which perch was used on a particular map. Nevertheless, it was probably usually the statute measure. The yard has always been a standard length and does not lead to complications.

In all the manuscript maps by both Christopher and Robert Saxton conventional orientation is observed, north being uppermost, with the cardinal points in either Latin or English written in the border (excepting Gomersal).

The borders are comprised of widely spaced parallel lines on all Christopher Saxton's maps, excepting the two of land in Kent where they are of two sets of narrow lines enclosing a wider space. Robert Saxton's borders are in the former style, excepting Gomersal which has no border. The cardinal points, centrally placed on each side of the map, within the border, are all in Latin on Christopher Saxton's maps until 1598, with the exception of Priestthorpe, and from thenceforth are in English with the sole exception of Snapethorpe, which is in Latin. Robert Saxton's are all written in English. On only one original map, Snapethorpe, is there a compass rose and that as an additional feature; Ashwell also has one, but this was probably added by the copyist.

The title cartouches are all of similar style, being simple double line boxes with slight variations, with only three exceptions – the two maps of Kent and that of Snapethorpe. Some have an additional narrow line to the inner 'frame', some have these additional lines to the top and right sides only, and some have diagonal lines joining the corners of the outer and inner 'frames'. All Robert Saxton's title cartouches are the same widely spaced double line boxes with additional narrow lines to the top and right sides of the inner 'frame', as portrayed in the map of Baildon (Plate 17).

On fourteen of Christopher Saxton's maps he employs a scale of sixteen perches to one inch; on one, twenty perches to one inch; on three, forty perches to one inch; on one, an eighth of a mile to one inch, and on one, half a mile to one inch. No scale is given on the map of Ribston and it is missing on the map of Thornhill. Robert Saxton used scales of sixteen, twenty and forty perches to one inch. Both men generally indicated the scale by a measured bar, often surmounted by a pair of compass dividers, and with a verbal statement of its representation in perches.

Geographical features are depicted in similar style to the engraved county maps, with hills drawn symbolically, resembling 'sugar loaves', generally shaded on the western slopes in contradiction to the engraved maps, where the shading was generally on the south-

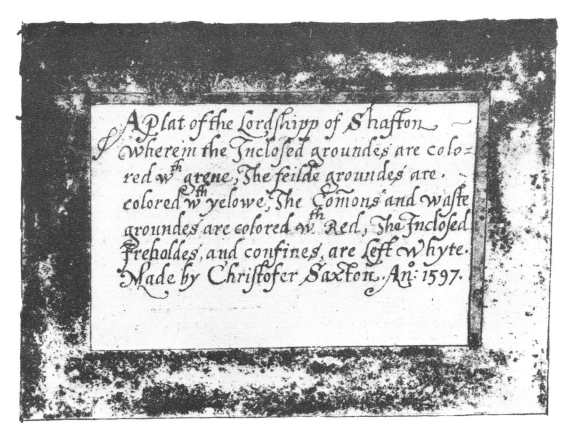

Typical title cartouche, taken from Christopher Saxton's map of Shafton.

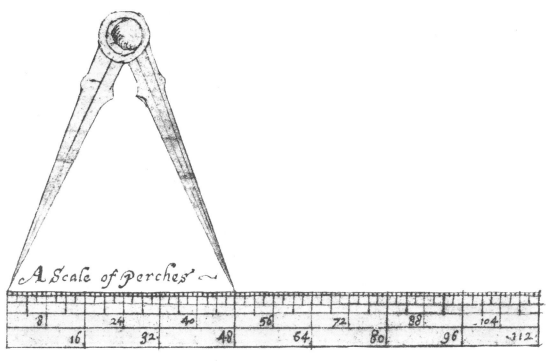

Typical scale bar surmounted by open compass dividers, taken from Christopher Saxton's map of Shafton.

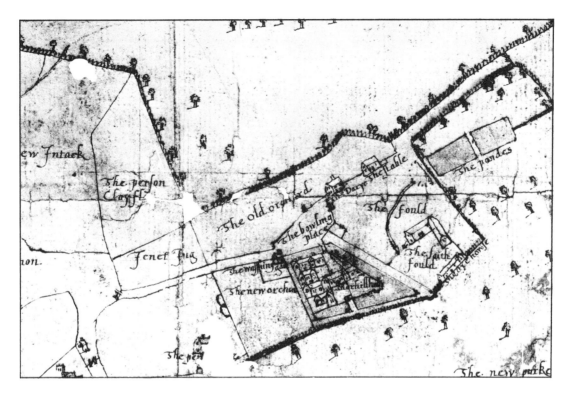

Thornhill Hall and surrounding area showing several typical features, taken from Christopher Saxton's map of Thornhill

eastern slopes. Watercourses are again shown as closely drawn parallel lines, often coloured blue, with bridges drawn as back to back convex lines, in much the same way as now used by the Ordnance Survey. Woodland and trees are shown by conventional tree-symbols and were often roughly coloured green, with shadows falling to the west. Parkland is identified by ring fencing, or paling, as illustrated in the map of Elland (Plate 13). Buildings are defined, with, in general, little distinction between houses; they are roughly drawn in pictorial perspective as a box, usually presenting two sides to the observer, with dashes representing doors and windows and with sloping roofs. The symbol used to signify a windmill is almost identical to that used by the Ordnance Survey today for a windmill, except that Saxton attaches the sails to the mill. Mansions and churches are more elaborately drawn and may have been intended to portray the actual building, though no work has been carried out to ascertain their correctness. Larger buildings are almost always named, or the occupant's name given, as generally being of significance to the case for which the map was made.

The information given on any one map is rather sparse, and only necessary features included, no attempt being made to produce attractive maps or fill empty spaces as on the county maps. They were produced for a purpose and irrelevant information was excluded. The result is a wide variety of maps from some giving minimum detail, such as that of Horbury which merely shows the course of the river, to those of whole manors, such as Aveley and Shafton, which appear to include every feature.

Saxton had a wide variety of employers at this stage in his career from the Duchy of Lancaster for the crown and other courts investigating disputed lands, boundaries or water rights, to institutions such as St Thomas' Hospital and private landowners who required their own estates surveying. A connection can often be traced between private landowners who employed Saxton: for instance he was engaged, at different times by George Savile, Edward Talbot and William Cavendish who were all connected by marriage. William Cavendish of Chatsworth was the second son of Bess by her second marriage to Sir William Cavendish. William's sister Mary married Gilbert Talbot, 7th Earl of Shrewsbury, and son of Bess's fourth husband George Talbot, the 6th Earl. Gilbert Talbot's brother Edward became the 8th Earl and their sister Mary married Sir George Savile. William Cavendish's elder brother Henry married Grace Talbot, sister of Gilbert and Edward. One can imagine a family gathering at which the expertise of Mr Saxton was discussed; at that date it was uncommon to have one's lands mapped and surveyed, there must have been some excitement and speculation at first seeing one's property mapped to scale on paper.

Two records of payment to Saxton survive, his remuneration whilst engaged by St Thomas' which was at 6s 8d per day, and a lump sum for his work for the Earl of Northumberland.

When Saxton wrote his own name he invariably spelt his first name Christofer, unless abbreviated to Christ:, or Christo:. The only times it was spelt Christopher was when written by others as in depositions or on later copies of his work, such as the map of Ashwell.

Robert Saxton has been largely neglected in the past, with only two examples of his work, Manningham and Baildon, discussed in detail. From surviving evidence it appears that he never sought employment outside Yorkshire. Robert Saxton is referred to throughout by his full name; the use of the surname only refers to Christopher Saxton.

The details of both Christopher and Robert Saxton's manuscript maps and written surveys follow. Where Christopher Saxton undertook more than one commission for a particular employer, these have been grouped together to give a more comprehensive picture. For instance, work undertaken for St Thomas' Hospital, the Savile family, Edward Talbot, Michael Wentworth, Richard Goodricke, William Cavendish and the Earl of Northumberland is described in chronological order under the date of the first known work for each employer. Single commissions and work undertaken in connection with legal disputes are described individually in chronological order (with the exception of the two maps of Wadsworth which are described together). A complete list of all Christopher Saxton's known commissions from 1587 to 1608, and Robert Saxton's commissions from 1607 to 1619, is given in chronological order in Tables 9 and 10, pages 79 and 122.

Measurements of maps are given for the whole sheet unless otherwise stated and are to the nearest half centimetre. Vertical measurements precede horizontal. All maps are on paper unless otherwise noted. Old style dates have been used where appropriate. The new year commenced on March 25th and the period from January 1st to March 24th is dated the year previous to that commencing on March 25th.

Chapter 8

Christopher Saxton's manuscript maps and surveys

TABLE 9

Christopher Saxton's manuscript maps and surveys

Date	Place	Map or Survey	Location	Page
1587	BROUGHTON, Yorks.	—	Public Record Office (Evidence only)	80
1587	CROFTON, Kent	M & S	Greater London Record Office (Evidence only)	80
1588	HASTINGLEIGH, ALDCLOSE, COMBE, COMBE GROVE, FANSCOMBE, Kent	—	Greater London Record Office (Evidence only)	80
1588	TEVERSHAM and FULBOURN, Cambs.	S	Folger Shakespeare Library	80
1589–90	GERONS, TAYLERFERES STEWARD, Essex	—	Greater London Record Office (Unattributed, evidence only)	80
1590	MONK BRETTON, Yorks.	M	Public Record Office (Evidence only)	85
1590	BAYFORD and GOODMANSTERNE, Kent	M	British Museum	85
1590	FAVERSHAM, Kent	M	Kent County Archives Office	85
1591	MEDLOW	S	Public Record Office	87
1591	HEATON COMMON, Yorks.	—	Evidence only	88
1592	PRIESTTHORPE, Yorks.	M	Public Record Office	88
1592	HINTON NETHERHALL, Cambs.	S	Greater London Record Office	80
1592	METHLEY, Yorks.	S	Original lost, transcribed by Fordham 1927	89
1592	RABBES FARM, Middx.	—	Greater London Record Office (Evidence only)	80
1593	AVELEY, Essex	M	Greater London Record Office	80
1594	HINTON, Suffolk	M	Original lost, copy at Suffolk Record Office	91
1594	WADSWORTH, Yorks.	M	Original lost, copy at Nottinghamshire Record Office	93
1596	ISLE OF AXHOLME, Lincs./Notts.	M	Public Record Office	99
1596	ACKWORTH, Yorks.	S	Pierpont Morgan Library	101
1596	MANCHESTER, Lancs.	—	Evidence only	100
1596–7	WHITCHURCH, Salop	S	Salop Record Office	101
1597	SHAFTON, Yorks.	M	Privately owned: Osberton Hall	101
1597	ELLAND, Yorks.	M	Nottinghamshire Record Office	93
1598	HORBURY and NETHERTON, Yorks.	M	Public Record Office	102
1598	OLD BYLAND, Yorks.	M	Public Record Office	105
1599	HUNSWORTH and EAST BIERLEY, Yorks.	M	Nottinghamshire Record Office	93
1599	LUDDENDEN, Yorks.	M	Public Record Office	106
1599	NOTTON, Yorks.	M	Wakefield City Museum	108
1599	WOOLLEY, Yorks.	S	Brotherton Library	108
1599	RIBSTON and WALSHFORD, Yorks.	M	Early copy. Privately owned: Ribston Hall	109
1600	WALTON HEAD, Yorks.	S	Leeds City Archives	109
1600	DEWSBURY, Yorks.	M	Dewsbury Public Library	111
1600	HUNSINGORE and GT. CATTAL, Yorks.	M	Early copy. Privately owned: Ribston Hall	109
1600	HADDLESEY, Yorks.	S	Privately owned: Chatsworth House	112
1601	LUDDENDEN, Yorks.	M	Public Record Office	106

Date	Place	Map or Survey	Location	Page
1601	SNAPETHORPE, Yorks.	M	Privately owned	114
1602	THORNHILL, Yorks.	M	Nottinghamshire Record Office	93
1602	THORNHILL, Yorks.	S	Nottinghamshire Record Office	93
1602	THORNHILL, EMLEY, ELLAND, SKELMANTHORPE, STAINLAND, MIDDLESTOWN, BOTHAM, RUSHWORTH, BARKISLAND, RAWTENSTALL, WADSWORTH, etc., Yorks.	S	Privately owned: Savile Estate Office	93
1602	BURLEY, Yorks.	S	Yorkshire Archaeological Society	115
1602	WADSWORTH, Yorks.	M	Original lost, copy at Nottinghamshire Record Office	93
1604	KEIGHLEY, Yorks.	S	Privately owned: Chatsworth House	112
1604	ASHWELL, Rutland	M	Original lost, copy North Yorkshire Record Office	115
1605?	COLD HIENDLEY, Yorks.	S	Brotherton Library	108
1606	RIVER SWINDEN, Lancs.	M	Public Record Office	118
1606	SPOFFORTH, Yorks.	M	Privately owned: Alnwick Castle	116
1608	SPOFFORTH, Yorks.	M	Privately owned: Alnwick Castle	116
1608	LINTON, Yorks.	—	Evidence only, Alnwick Castle	116
1608	LIVERSEDGE, Yorks.	—	Evidence only	119

Broughton, Yorkshire

The first evidence of Christopher Saxton's subsequent employment following his national survey and production of the map of England and Wales is found in a report made to the Exchequer Court in May 1587.[1] A Special Commission of Inquiry had been appointed to survey, establish and measure the boundaries of part of the manor of Broughton-in-Elsack, three miles west of Skipton in the West Riding of Yorkshire, which belonged to the Queen and was late part of the dissolved monastery at Pontefract.[2]

The Commissioners had given fourteen days notice of their intention to meet at Broughton Church on March 20th, 1586, but the document had been scornfully flung to a spaniel dog and sent back after the messenger. Henry Tempest and his two sons, presumably Stephen and Henry, were adamant in their refusal to give evidence to the Commissioners, who had been accompanied by Saxton to make the survey. The town bailiff, being a tenant of Mr Tempest, failed to appear, and locals who directed the Commissioners were threatened. The Commissioners and Saxton were 'rescued by two of Mr Tempest his sonnes', implying in twentieth century colloquialism, 'seen off'. Further arrangements were made to meet at Wetherby a fortnight later to examine witnesses, but none appeared and the matter was reported by the Commissioners to Lord Burghley, Sir Walter Mildmaye, Sir Roger Manwood and the rest of the Barons. Depositions were eventually heard at Skipton in November 1587.[3]

Added at the bottom of the Commissioners' report is a note: 'm^d that I was p^{r}sent, helping Marmaduke Wilson when we were rescued by Mr Tempest his two sonnes in the Survey' signed 'Christ: Saxton'. The signature is obviously Saxton's, but the statement is not in his hand.

Crofton, Kent; Hastingleigh, Aldclose, Combe Grove, Fanscombe and Comb, Kent; Fulbourn and Teversham, Cambridgeshire; Gerons, Talerferes, Steward, Essex; Hinton Netherhall, Cambridgeshire; Rabbes Farm, Middlesex; Aveley, Essex

But for the fact that Christopher Saxton was involved in the abortive survey at Broughton in Yorkshire, it would have appeared that he remained in the south of England for some years after completion of his national works, but this was obviously not the case, although he undertook the majority of his commissions in the south of the country during the next few years. These southern commissions were, however, interspersed with work in the north of England and Saxton must have travelled north and south several times during this period.

Saxton was employed by the Governors of St Thomas' Hospital in Southwark, London, on several occasions to survey their country estates. References to his employment appear in the Minute Book kept of the meetings of the Governors of the hospital,[4] and records of payments made to Saxton are found in the Rental or account book in which financial details, tenants' rents and expenses incurred were entered.[5] That Saxton was employed on a regular basis is indicated by the fact that he is referred to as 'our surveyor' in the Minute Book, without stating his name, with a later entry in the Rental that Saxton was paid for the survey in question, confirming that he had indeed undertaken the work. Unfortunately only one map, that of Aveley in Essex, dated 1593, and two written surveys of lands in Cambridgeshire, dated 1589 and 1592, survive, but it is clear that Saxton was also employed to 'survey and plott' the manor of Crofton in Kent in 1587, followed by other land in Kent in 1588 and a farm in Middlesex in 1591–2. There is evidence that Saxton was employed by the Governors of St Thomas' Hospital during the period 1587 to 1593, but in fact his work may have extended over a longer period as not all details of surveying arrangements are entered in the Minute Books. For instance, in the case of the survey of Crofton in Kent, the only entry is on November 13th, 1587, when Saxton attended the court and displayed his 'survay and a plott' which he had lately made. There is no earlier entry of a decision to engage Saxton to carry out the work, confirming the opinion that the minutes do not give a complete picture of his work for the Governors. It seems likely, however, that the survey of Crofton was his first commission as the entry included the instructions 'that for that paines & travell wch he hath taken & wch he shall hereafter take about the like busynes for the surveyinge of any of the lands belonging to this house he shall have paid unto him the some of Six shillings & Eight pence a day', and also that the Clerk, Lambert Osboltson should accompany Saxton and assist him 'from tyme to tyme'. Saxton actually took six days to complete the survey of Crofton and was paid forty shillings for the work. Payments in units of 6s 8d or 13s 4d are commonly found in the account books of the hospital; it was a convenient unit of currency at the time, corresponding to the value of the coins in use – the mark, which was worth 13s 4d, and the noble, or half mark, worth 6s 8d.

St Thomas' Hospital had a very early foundation, its history can be traced from the twelfth century, but in 1540 the hospital was surrendered to Henry VIII and was closed until 1552. In 1551 St Thomas' Hospital and what could be salvaged of its former estates was sold to the City of London. The estates of the Savoy Hospital, which had also closed, were made over to the City by Edward V/ and earmarked for the maintenance of St Thomas' Hospital. The Savoy Hospital owned a considerable amount of land both in London and the country, and it was these former estates of the Savoy Hospital which Saxton was employed to survey some years later. As St Thomas' was owned by the City of London, the Court of Governors of the hospital was confined to the Aldermen and Common Councilmen of the City, and the names of some notable and influential gentlemen appear as Governors present at the fortnightly meetings held to discuss administration of the hospital. There was a considerable amount of property held in London itself, but although there are entries in the Minute Book of a surveyor being required to examine property in London, no specific mention of Saxton is found in connection with these local holdings.

It may be assumed that by the mid 1580s Saxton's name at least, and possibly he personally, would be familiar to the City fathers, when he must have been accepted as the leading surveyor, and the establishment of this connection with the gentlemen of the City must have stood him in good stead when employment was obtained through personal recommendation.

As already noted, Saxton submitted a survey and a map of the manor of Crofton to the Governors of the hospital in November 1587. Neither of these survive today in the archives of the hospital. Crofton Manor, near Orpington in Kent, was leased by John Staple in 1587. It later became Crofton Pound Farm, with 245 acres, and was still held by St Thomas' in the 1930s.[6]

The following year Saxton was employed to survey other manors in Kent belonging to the hospital, and an entry in the Minute Book on April 22nd, 1588, reads: 'Also it is ordered that Saxton the Surveyor shall goe forward to Survey or mannrs in Kente as his leysure will serve him.' A later entry in the Rental reads: 'Item paid to Xpofer Saxton or Surveyor for xxj Daies travel about the Surveying of ye Mannrs of Hastingleye, Aldclose, Combe, Ffanscombe & Grove, & our farme at Comber at vjs viijd a day – viil.' These manors are known today as Hastingleigh, Aldclose, Combe Grove, Fanscombe and Combe, which are all near Wye in Kent. Neither a map nor survey of these manors in Kent survives today in the hospital archives, and there is no record of any modern writer having seen them.

The manor of Hastingleigh and Aldclose was held by Robert Allison in the late sixteenth century. His lease, granted in 1583, is characteristic of others and includes the stipulation that he is to keep continual hospitality upon the premises. It is probable that Saxton and Lambert Osboltson would receive hospitality from the chief tenants of the hospital during their surveying of the estates. It was certainly customary for the tenants to accommodate the Governors when they visited the estates, either for an inspection or to hold manorial courts.

The surveying of these various manors near Wye took Saxton twenty-one days, and at the rate of 6s 8d a day he received £7. This was a high rate of pay when compared with other basic wages paid to the hospital staff. For example, the Hospitalier or chaplain and the clerk both received £10 per annum, the cook £8, and the surgeons £15 each. The resident staff, however, received 'perks' in addition to their wages so that a clear comparison is difficult to make.

There are records in the Minute Book and Rental of two surveys carried out in 1589, but Saxton's name is not mentioned in connection with either. The first reference, in a minute dated May 15th, 1589, reads: 'It is ordered that our Lands at Ffulborne & Towsham wch Mr Wise holdeth by lease shalbe surveyed betwene this & Bartholmewtide, and that there shalbe Courte kept in Cambridgshire between Barthewtide & Michaelms.' In the annual accounts it is recorded that there was '... paid for all manner of charge touchinge the Survey of the land in Tevershm & Fulborne in Cambridshire as by the pticulers appeareth ... xls xvijs' which would mean that the surveyor was engaged for over thirty-two days. Fulbourn and Teversham are situated to the east of Cambridge.

These surveys have recently been discovered in America and are indeed by Christopher Saxton.[7] The two are stitched together in a booklet containing twenty-four pages of text in an unknown hand. It is endorsed: 'A true Terrour and contente of the landes in Teversham and ffulborne in the County of Cambridge.' The first survey, which is eight pages long, is headed: 'A true Terrour and contente of all the Lands called Manners and Allyns in Tevershm wthin the County of Cambridge fomely ... pcell of the Possessions of the Savoie, & now belonging to St Thomas Hospitall in Southwerk mad by Christofer

Saxton in Julie A⁰ 1589.' Measurements are given in acres, roods and perches only. Saxton firstly describes the enclosed land, which totalled 27a or 2p. He then details lands lying in the common fields, Townesendefielde, Holmefielde, Brookefielde, the Churchcrofte and Myllfielde. The total content of land surveyed in Teversham was 125a 1r 23p. The location of each field strip is described, but no tenants named.

The second survey, which covers eighteen pages, is headed: 'A Terrour and contente of all the Lande in the parish of ffulborne in the County of Cambridge formely in pcell of the possessions of the Savoy & now belonging to St Thomas hospitall in Southwerk made by Xpofer Saxton in Julie 1589'. Again the enclosed land, totalling 16a or 16p is described first, followed by 31a 1r 0p comprising the Hospital Heath. Saxton then describes the contents of the four common fields, Woodebridgefielde, Crossefielde, Highfielde and Smallwayfielde, in similar detail. For example, in the Crossefield, 'Item 1 pece in Greneway furlonge betwixte Peterhowse North & Ware South – 1r 30p. One pece in the same furlonge betwixte Wright Sowth & Quenes colledge north – 2r 4p.' The total content of the lands surveyed in Fulbourn was 262a 3r 25½p, most of which is set out in quantities of less than two roods.

The second unattributed survey recorded in the Rental for the year 1589–90 was of the manor of Gerons, Taylerferes and Steward and other land in Great Parndon, in Essex, for which £9 6s 6d was paid, a total of twenty-eight days work. No corresponding entry appears in the Minute Book, but as with the survey of Fulbourne and Teversham it seems very likely that it was carried out by Saxton. The survey does not survive in the St Thomas' Hospital archives, nor amongst the manorial records deposited by the Governors in the Essex Record Office. Great Parndon is near the Hertfordshire boundary with Essex. It was at this same period, on April 14th, 1589, that Lambert Osboltson, the Clerk of St Thomas', approached the Governors to consider whether he should be paid 'for his paynes & travell to be taken in the Surveying of o⁰ lands, for that his to goe wᵗʰ Mʳ Saxton the Sʳveyʳ & to attend upon him, when he is about the same'.

The next record of Saxton's employment by St Thomas' Hospital is in 1592 when he made a written survey of Hinton Netherhall near Cambridge. An entry in the Minute Book for May 5th, 1592, reads: 'At this Court It is ordered that there shall be Court kept in Essex and Cambridgeshire about the end of next terme, And that there shallbe a Survey made of the Lande at Hinton before the Courte be kept there, yf the Surveyor can possiblie be there so soone.' It will be observed that no mention is made here of Saxton by name, but a written survey of Hinton Netherhall dated 1592 and signed by Saxton, survives in the hospital archives.[8] To complete the picture, an account was entered of Saxton, 'our Surveyor', having been paid iiij¹ xiijˢ iiijᵈ (£4 13s 4d) for fourteen days' work in surveying Monnsaugh's farme in Middlesex and Hinton in Cambridgeshire. The survey of Monnsaughes farm has not survived; it was actually a farm called Rabbes occupied by William Monnsaugh.

The survey of Hinton is written on three sheets, but unfortunately the top left corner of the first page is torn off and part of the title is lost. The remaining portion reads: '... Mañor of Netherhall in the pishe of Hynton ... of Cambridge. Belonging to St. Tho: Hospitall ... of Willm̄ Catlyn gent: and others, Made ... er Saxton in Julye An⁰. Dni / 1592 / 35 Eliz:'. Inscribed on the back is 'A Survey or Terrar of the Demesnes of Hinton Netherhall Taken ... Mʳ: Saxton A⁰: 1592'.

The survey itself is written by Saxton, but the endorsement is not. The land measurements, recorded in acres, roods, dayworks and perches, show that Saxton surveyed a total of 187a 1r 2dw 3½p. The writing on the survey is clear and legible, but the numerical additions are not always correct. The survey basically details the site of the manor called

Netherhall Close with the orchard, etc., and adjacent fields (Fendon, Quarrye, Bridge, Heath, Church and Yonnton Fields). It is now bound up with other surveys of St Thomas' Hospital lands in the hospital archives.

The final known commission undertaken by Saxton for St Thomas' Hospital was to map the manor of Aveley in Essex which lies close to Thurrocks and Purfleet on the north bank of the River Thames. The map, which survives in the archives of the hospital,[9] has the title in the top left corner: 'A Platt of the Demesne Lands of the Mannor of Alvethley in the Countie of Essex, which John Herdson Esquoir holdeth of Sainte Thomas H(os)pit(al) by Lease; Wherein all the Marshe Landes are coloured with yellowe and uplande groundes wth redd, and the confynes lefte white; wch platt was ma(de b)y Christofer Saxton in June A.. Dni 1593.' The map measures 80½ cm by 128½ cm and is on paper, now mounted on cloth, glazed and framed.

A scale rule in the bottom right corner measures three inches, is coloured red and divided into perches, being marked at 8, 16, 24, 32, 40 and 48 perch intervals. Above is the statement 'This Scale contayneth 48 Perches', giving a scale of sixteen perches to one inch. The paper is badly worn, particularly around the edges and where the map has previously been folded. The map is drawn by Saxton, but has additional notes and tables at the bottom in an unknown hand. It is probable too that the acreages given in the fields, described in acres, roods and perches only, are not in Saxton's hand. Some of the original colouring survives, the roofs of most of the buildings are painted bright red, and the yellow outline round the marshland is still distinguishable, but the red outline round the 'upland groundes' has faded beyond recognition. In the bottom left corner of the map on the River Thames are two crudely drawn ships. One is merely a blocked in shape, presumably representing a hulk; the second has suggestions of a mast and rigging. Apart from one wood, no trees are drawn on the map, and only one hill, at Purfleet, is shown, although the upland ground is shown as an area.

The map shows the River Thames in the south-west from Wellington Creke to Purfleet, with reed beds, marshes and a marsh-wall. Purfleet itself consists of two houses and a windmill on a hill, with 'Purflete mylne', with a waterwheel, on the estuary of Mai Dike (unnamed). The marshland extends for a considerable distance inland to a hamlet called Marshfoote, with 'The way from Marshfoote howse to Aluethlay', drawn as double lines, leading to the north-west. 'Aluethlay Towne' is shown as a long street lined with buildings, with three detached buildings in the middle of the road. To the south of the main street is the church, 'The Lords pound', and an enclosure wherein is written 'yt ys said that the maner howse of Alverlay stood in this place'. In the middle of a road leading south from Aveley are two small rectangles, 'the Buttes'. All fields are named, and most tenants.

The survey of Aveley took Saxton just over twenty days to accomplish for which he was paid £6 16s 8d at the customary rate. In the Minute Book is an entry, in November 1589, recording that John Herdson was a suitor for the confirmation of the lease of the manor of Aveley which he claimed he had lately bought from the Earl of Oxenford. The lease was confirmed to Herdson in December 1589.

A copy of Saxton's map of Aveley is to be found at the Essex Record Office.[10] It bears the title (which differs from the original): 'A Plan of the Demesne Lands of the Mannor of ALVETHLEY in the County of Essex belonging S^t: Thomas's Hospital wherein the Uplands are Coloured with Red, the Marsh Lands Green, the Salts and Reeds with Yellow (being a true Copy of Christopher Saxton's Survey in 1593) by J. Freeman 1782'. The cartouche is composed of scrolls and acanthus leaves, in late eighteenth century style. The copy has lost its cardinal points in the margin and these are replaced by an eight point compass indicator in the centre bottom of the map. The script is that of the copyist and does not attempt to reproduce Saxton's hand. The additional notes on Saxton's original map are reproduced on the copy. The boats on the Thames are missing, but basically the map content is true in the copy. According to Eden, a Joseph Freeman was working as a surveyor in the late eighteenth century in East Anglia, and was at one time Steward to St Thomas' Hospital, which would account for his having access to the original map.[11]

The mapping of Aveley was, as far as is known, the final task undertaken by Saxton for St Thomas' Hospital. Why this regular employment ceased is not indicated. Perhaps all the necessary surveys of hospital lands had been completed for the time being; there are several entries in the minutes which record that certain holdings 'shall (be) specially surveied at the next genall viewe of o^r Lande'.

One final note worth recording, of interest to cartographic students, is an entry in the Minute Book on January 23rd, 1585, which reads: 'Item it is ordered that Mr Robert Tailor shalbe one of the Surveyors in liewe of Mr Bate who by reason of sickness is unable to travell in the affaires of this house.' Presumably Saxton's predecessors.

Monkbretton, Yorkshire

In 1590 Saxton was requested to map the manor of Monk Bretton to the north-east of Barnsley in the West Riding of Yorkshire, thirteen miles south-east of Dunningley.

After the dissolution the manor of Monk Bretton remained in the possession of the crown until the time of James I. In 1590 William Foster, on behalf of the Queen's tenants, petitioned Lord Burghley to compel Richard Wortley to lay open the grounds he and his undertenants had enclosed. Foster also requested that Christopher Saxton, the Queen's surveyor, make a plot of the manor and that a commission be established to make enquiries. With the petition is an order, dated July 16th, 1590, for Saxton and Charles Jackson, steward of the manor of Monk Bretton, to make a 'trew platte' of the land in controversy.[12]

Subsequently a commission was set up, but the map does not survive with other documents at the Public Record Office and must be considered lost.[13]

Bayford and Goodmansterne, Kent; Faversham, Kent

Christopher Saxton was employed privately to map two estates in Kent; both maps bear Saxton's name and both are dated September 1590. It cannot be coincidence that these two maps were prepared in the same month, but as yet the connection between them, probably a connection between the owners of the lands mapped, has not been established. The two estates were certainly near each other, but not contiguous. Both maps appear to be estate surveys, and are executed in similar style with rather more embellishment than Saxton's other maps. For instance, both have the same elaborate design of title cartouche resembling carved wooden scrolls surrounding a box frame in the Italian style, similar to those found on many contemporary printed maps of the period. Both maps have a detailed neatly drawn pair of open dividers surmounting the line scale, the legs of which are identical on both maps, but the tops differ slightly in design. Both have the verbal statement over the line scale 'This Scale conteyneth 64 perches', with a line rule measuring four inches, giving a scale of sixteen perches to one inch. Both maps have the same border which differs from that found on Saxton's other maps (see page 75); and finally both maps are drawn on parchment which again is unusual.

The map of Bayford and Goodmansterne is in the British Museum.[14] It measures 74 cm by 65 cm. The title, in the top right corner, reads: 'A Plate of the Mañers of Bayford and

Goodmanston in the parish of Sittingborne in the Countie of Kent. And made by Christofer Saxton in the month of September. An⁰ Dm̄i. 1590.' Beneath the title, and almost double the size, is a summarized list of the areas surveyed. It is headed 'The content of the Manors of Bayford and Godmanston' and lists 'The Scites of yᵉ Mañors' giving brief details and contents, in acres, roods, dayworks and perches, of the 'Arable Lands', the 'Middowe' (meadow), the 'Fresh marshe', and the 'Sault marshe'. The 'Som totalis' entered is 196a 2r 6dw 0½p, which a quick check reveals inaccurate, on this occasion by 1a 3r 2dw. However, a cross check from the reference table with the contents of the lands marked on the map also reveals discrepancies; they can most probably be attributed to carelessness in copying. The contents' summary is contained in a simple line box in the same style as the border.

In the bottom left-hand corner of the map is the scale rule surmounted by dividers. The rule is more elaborate and detailed than was Saxton's usual custom, and is divided horizontally into five sections, in addition to the usual vertical divisions. It is divided and marked at sixteen-perch intervals at the bottom, into eight-perch divisions above, and so on until the top line is divided into sixty-four perches. Only the bottom divisions are numbered.

The land mapped takes up less than half the surface of the parchment and almost appears to be cramped into the left side of the map. The road through Sittingbourne in the south-west is lined with houses which differ slightly in style from Saxton's usual 'boxes', many having sloping roofs with the ends of the house drawn as an inverted V, suggesting timber-frame construction. There is no indication of high ground. The road is marked 'the waye from London to Canterburie', and runs due west east across the bottom of the map. The river, in the north-west corner of the map is named Miton Creke and is bounded by salt marshes. Four ships are shown at the entrance to 'Sittingborne Creke', drawn in the same style as those on Saxton's map of Aveley, with crude rigging surmounted by a flag. All the fields are named and most have their acreage given; in Swantston field is a note 'pte of yᵉ Quenes Land' and a portion of the field is enclosed and has the note 'John Carters two acres which he would sel yᵉ Lord'. The sites of the manors of 'Goodmanston' and Bayford are both marked.

The map itself is in good condition and complete, but some of the writing has become obscured along the fold lines. The map was coloured in the usual manner, the cartouches, line scale, roofs, trees, water and roads being painted as customary, but with the boundaries of the estate painted pink.

There is some doubt as to whether Saxton drew this map himself, and even after close comparison with others indisputedly by him it is difficult to be certain either way. It is clear that this and the Faversham map are by the same draughtsman, and whilst having more elaborate title cartouches, are basically in Saxton's style. The writing, at first glance, appears to differ in certain respects, but again on close comparison with his known hand shows many similarities. The portrayal of the ships, for instance, is identical to that on the map of Aveley drawn in 1593. One must perhaps make allowances as these maps were drawn on parchment, as opposed to the usual medium of paper, which might result in occasional variations. The question of attribution must remain open.

The manors of Goodmansterne and Bayford were situated at Sittingbourne and Milton on a tributary of the River Swale which flows into the Thames in north Kent, to the south of the Isle of Sheppey. The manors were sold by Thomas Finch to Sir William Garrard, who had been Lord Mayor of London in 1555.[15] The estate descended to his grandson, Sir John Garrard, of Whethamsted in Hertfordshire, who was Treasurer and one of the Governors of St Thomas' Hospital when Saxton was employed by that body. In 1601

Garrard became Lord Mayor of the City of London, having been Sheriff 1592–3. The map was presumably a survey of his estates.

The second map of Kent, of Homestall Farm at Faversham, is deposited at the Kent County Archives Office.[15] The title, in the top right corner, reads: 'A plat of Homston Farme in the parishe of Feuershm̄ in the Countie of Kent and in the tenor of Henrye Saker, which plate is colored with red, that which is colored with Yelow belongeth to th Mañer of westwood, and ye confines are left whit. Made by Christofer Saxton, in September, Anº: 1590.' Immediately beneath the title is the scale rule, which has decorative arrows half way along each of the four sides, with dividers above. The rule is marked in the same way as on the Bayford map. The complete map measures $72\frac{1}{2}$ cm by $65\frac{1}{2}$ cm.

To the bottom left of the map is a rectangular box with double inner and outer lines, and a panel at the top decorated with scrolls in which is written in gothic script: 'The content of Homston farme'. The contents listed below, given in acres, roods, dayworks and perches, are divided into Arable Land and Pasture, which together with the site of Homston Farm with its orchard, hopyard and garden gives a total of 290a 2r 3dw 3p surveyed. A check of the additions reveals the total to be two perches short and again some of the measurements given in the fields do not correspond with those in the summary table. Another example of the inaccuracy we come to expect.

The land surveyed is contained in the top left and bottom right quarters of the map and is bisected horizontally by 'Homston Lane Leading from Feuershm̄ to Hernhill'. To the south the land stretches to the 'London waye' and shows Homston Farm, a detached building with a pond surrounded completely by a range of buildings and a fence. Fields belonging to the farm are named and acreages given. Some adjoining landowners are named, notably Mr William Garrard. The upper portion of the map shows more fields and some marshland belonging to the farm, together with land owned by the Abbey of Faversham; it is bound to the west by Claygaite Lane. Ewell Creke is shown in the northwest corner. A note describes a disagreement over a right of way: 'Note that the farmer of Homston challengeth this waye to his close called the 14 acres as wrong through the Scole land.'

The map is coloured as usual: water in blue, trees green, the farm roofs grey, and with the boundaries painted with a black line, often with a rim of yellow alongside. There is no indication of topographical relief. The map is endorsed 'HOMSTEER' and 'Mapp of the Farme of Homstede'. It is in good condition, apart from slight crumpling at the top left, and is entirely legible. The writing is in the same hand as the map of Bayford and Goodmansterne, and the same remarks apply concerning attribution.

Medlow

A hitherto unpublished written survey by Christopher Saxton was recently discovered at the Public Record Office, London.[16] It probably lay unrecognised due to the fact that the area surveyed cannot be located today.

The survey is on a single sheet and headed, in an unknown hand, 'The Suruay of Medlow, made by Christofer Saxton, Anº: 1591'. It is comparatively short and divided into two sections; firstly the fields, closes, meadows, etc. are listed, and secondly, the woods. The former includes lands with such distinctive names as Philpot Close, Batesworth field and meadow, Jackson's Laund and Farwells Close, which, with others – totalling eighteen – measure 686a 1r 7dw 1p. The woods, of which fourteen are named, include Nether and Over Battesworth Woods and Pennyman's Wood, measure a total of 227a 1r 4dw 1p, giving a sum total surveyed of 913a 3r 1dw 2p. At the conclusion is the signature

'B me Christ: Saxton' in writing which one would not normally attribute to the surveyor.

Before inspection it was thought that this could possibly be the missing survey of Methley in West Yorkshire, named Medley by Saxton on his county map of Yorkshire, but despite the similarity of name this is definitely not so, as the field and wood names do not correspond with those in the survey of Methley transcribed by Darbyshire and Lumb (see page 89).

It may possibly relate to land in Midloe in Huntingdonshire, recorded as Medlowe in the sixteenth century, but this suggestion remains uncorroborated. The survey is in the Special Collections classification at the Public Record Office which includes papers from a number of sources, but this provides no indication of provenance.

Heaton Common, Yorkshire

In 1591 Christopher Saxton made either a map or survey of Heaton Common in the West Riding of Yorkshire, three miles north-west of the centre of Bradford and twelve miles from Dunningley. Whichever it was cannot now be ascertained as no trace of it can be found.

The only recorded evidence of the document is contained in a deposition concerning a dispute over a coal mine, recorded by Wilfred Robertshaw in a paper 'The Manor of Chellow'.[17] Robertshaw writes that in 1765 a chancery suit toured the boundaries between Chellow Common and Heaton Common (in Bradford-dale), in a dispute about a coal mine opened in Chellow Common. A bill of complaint was brought by Richard and William Hodgson, then the largest landowners in Chellow, against John Field, Lord of the Manor of Heaton. A William Bolling of Ilkley stated in his deposition that he had seen a map or survey of Heaton Common, made in the year 1591 by 'one Christopher Saxton amongst the documents belonging to William and Mary Thomas of Marylebone'.

No documents having belonged to the Thomases can now be traced. Mary Thomas, who with her husband, possessed a third of a moiety of Chellow, was the granddaughter and heir of William Bolling, who in turn was the great grandson of Edward Bolling of Chellow who died there in 1592. It was probable that Edward Bolling employed Saxton shortly before his death. Without the map or survey no comment can be made as to why it was made.

Priestthorpe, Yorkshire

In 1592 Saxton was again engaged to draw a map in his native county of Yorkshire. It was made in connection with disputed land and property in Priestthorpe in the parish of Bingley, in the West Riding of Yorkshire, some thirteen miles from Saxton's home at Dunningley.

The map is now at the Public Record Office, London.[18] The title, at the top right corner of the map, reads: 'The plat of A howse and certen Landes in controuersye in the pishe of Bynglay. called by Mr. Willm Ball plaintyffe, The psonage howse, and the glebe Landes thereto belonginge. And by Robart Waide defendant, The Mansion howse, or Manner place of Preestthorp, and the Demaines thereto belonginge. Which groundes are Colored with yelowe and the confines left white made by Christofer Saxton Anº: Dm: 1592.' Beneath the title, but within the cartouche, is 'The content is 56 acres 3 Roodes 2 Dawrkes 1 pche'. The cardinal points are in English instead of the customary Latin, with 'NORTH' missing, a portion of the paper being torn away. In the bottom left corner is the line scale rule, divided as on the maps of Kent, but with no dividers over. It measures two and a half inches and is marked at 8, 16, 24, 32 and 40 perch intervals, with the statement over: 'This Scale contayneth. 40 .Perches', giving an actual scale of sixteen perches to one inch. The map measures 44cm by 54cm, has been repaired and is mounted.

The disputed lands, shown in two blocks, are roughly colourwashed in yellow, with some roofs coloured a blue-grey. The inner section of the title box is carelessly coloured on the bottom and right side only in a light yellow-brown.

In the centre, at the south, is 'Bynglay towne' with the church at the western end; a row of houses on each side of a road comprises the rest of the town shown. The buildings all adjoin each other, but are set at different angles. The church itself may be symbolically drawn with its low tower, but does probably represent the appearance of it at that time – the upper portion of the tower not being added until 1737.

Priestthorpe and the disputed lands are shown a short distance to the north-east of Bingley, bounded by a road called 'Long Layne' to the north of the hamlet, which runs south into 'Prestethorp grene' and continues due south. 'Castle layne' is shown running from north to south – to Bingley – on the western side of the map. In the north-west is the 'West Feilde deuided into fowre closes', part of the disputed lands. Names of fields are given, but no land measurements shown. Adjoining landowners are named. The house which figured in the dispute is shown south of Priestthorpe Green and is simply called 'The howse'.

The legal case for which the map was prepared was an exceedingly complicated one, and is described at length by Dodd in a paper *Priestthorpe and the Rectory of Bingley*.[19] Dodd also reproduces Saxton's map of Priestthorpe. To greatly simplify the case, the property in Priestthorpe had belonged to Drax Priory until the dissolution in 1536. Successive vicars had lived in the house in question, but only as tenants, and by 1592, when the map was drawn by Saxton, the Vicarage was in the possession of Robert Wade. The Rectory tithes were leased by the Crown for life successively to William, Edward and Thomas Ball. William Ball had instigated legislation to extend a claim to include the Vicarage house, but his case failed. Papers concerning the case are also at the Public Record Office.[20]

Most of the rest of the land shown on Saxton's map was owned by Anthony Walker who became Lord of the Manor in 1596. His house is drawn on Saxton's map at 'Prestethorp grene', but the Walkers had moved into 'Gawthorp howse', also shown on the map, by the early seventeenth century when it was known as Gawthorpe Hall.[21]

Methley, Yorkshire

In 1592 Christopher Saxton was commissioned under a decree issued by the Archbishop of York and the Ecclesiastical Commissioners to survey the parish of Methley, which lies six miles due west of Dunningley, in the West Riding of Yorkshire.

The survey is quoted in full by Darbyshire and Lumb in their *History of Methley*, and they must obviously have had access to the survey.[22] Despite its existence at such a recent date it cannot now be traced, although searches have been made in the obvious locations, the Borthwick Institute, the Savile Estate Office at Methley, the Mexborough Archives at Leeds City Archives and with the Rector of Methley.

Darbyshire and Lumb head the survey 'The List of Persons and Property Liable to be Rated under the Archbishop of York's Decree of 1592. (By Christopher Saxton, Cartographer).' The words in brackets are obviously not on the survey; Saxton never described himself thus, nor spelt his christian name so.

The survey itself is set down in a rather different manner to Saxton's later surveys, as indeed it was made for a different purpose to any other. It is basically in two parts: the first being divided into two sections. Firstly, under the heading 'The Enclosed Grounds of Methelay' is written: 'The names of the owners, the nature of the ground, the particular contents, Acres, Roodes, Dawkes, Perches, the totall Summe Acres, Roodes, Dawks Perches.' Then follows a list of 103 persons, their property, land and acreage. John Savile, as owner of the manor, held the largest acreage of over 197 acres; only one other, Henry Farrar, held over 100 acres. A list headed 'The Feild Landes of Methelay' follows, which names fifty-nine men and their acreage in each, or some, of the five open fields. Concluding this first section is a note 'The totall some of all the lands in Methelay, the commons and wastes excepted, surveyed and measured by Christofer Saxton with the measure of 21 foote to the pearche. 1615 acres 3 roodes 0 dawkes 1½ pearches. Anno domini 1592. Per me Christ: Saxton. Note that the new rente of enclosier is paid after the rate of the les mesure viz. of xvi foote and a halff.'

The second part of the survey is a list of the landowners, the names of his fields and the type of ground, i.e. pasture, whinnie ground, etc., followed by the acreage and finally, his total acreage. This is, in fact, a detailed breakdown of the first section of the survey, the enclosed grounds, but the numerical additions are often incorrect, and the second detailed list does not always correspond with the first. As the second section has no descriptive heading it may be that other factors are included or excluded. As the original has not been inspected and the handwriting examined, it cannot be said whether the second section was indeed by Saxton, although this seems likely in view of the fact that dayworks are used. The fact that the two sections differ slightly in content may be due to their being taken at different times.

The whole survey concludes with a list of the commons and wastes, totalling 183a or 2dw 1p; the contents of the enclosed grounds, 1588a 3r 2dw 1p; the field grounds, 266a 3r 1p; and, 'Summa totalis after 21 foot to the Pearch 2038a 2r 5d 2p' (again slightly inaccurate), and the comment 'An acre measured by 21 foot to the Perch is an acre and an halfe and xix Perches and a halfe of the statute measure wch. is but xvi foot and a halfe to the Perch'.

In the early 1580s, John Young, a churchwarden of Methley, instituted a lawsuit against Henry Grice of Sandal and other parishioners, to enforce payment of a church rate for the repair of the church buildings. On July 30th, 1582, Edwin, Archbishop of York, and the Ecclesiastical Commissioners of the Province issued a decree that all the parishioners should contribute towards the repair of the church according to the amount of land each held within the parish. John Savile, owner of the manor, and Henry Farrar later reported that a rough survey of the parish was then undertaken by the inhabitants, some land being assessed by measurement, some by rough estimation and some omitted altogether. No distinction had been allowed between different qualities of land. This unfairness caused grievance, and on December 13th, 1591, the new Archbishop, John, and the Commissioners issued a decree from Cawood which required, at the expense of the parishioners, that they employ 'one Christofer Saxton or some other expert Surveyor, with their assistance to survey the said whole parish and to set downe everie messuage, cottage, acre and rood and lesser measure ... of arable, meadow, pasture and furres according to the statute or ordinance de terris mensurandis or by some other certein uniforme rate.'

Accordingly Christopher Saxton, together with Richard Shan and John Hagger the churchwardens, swore their oaths before the clerics and commenced a survey of the parish. For some reason the rate of twenty-one 'foote to the pearche or pole' was employed, rather than the more common computation or statute rate of sixteen and a half feet to the perch. The survey was presented to the commissioners in the great chamber of the Archbishop's castle at Cawood on December 19th, 1592, and after lengthy debating, rates were agreed upon whereby for every acre of main meadow or feeding ground a penny was payable, for every acre of less fruitful pasture, arable or woodland a half penny farthing, and for every acre of whinnie or furrey ground a half penny. Householders and cottagers were also assessed. The 'charges of the said Survey' together with expenses concerning the decrees, were to be borne by the parishioners out of the rate.[23]

Hinton, Suffolk

The Suffolk Record Office has a photograph of an unsigned map by Christopher Saxton of the Manor of Hinton in Blythburgh, Suffolk.[24] The copy was taken from the original owned by Sir Gervase Blois, since deceased, but which is now unfortunately lost, although it is hoped only temporarily.

The title of the map, in the top right corner, reads: 'A plat of the Mañer of Hynton in the pishe of Bliburgh. and Countye of Suffolke, Wherein all the demanes are colored wth yelowe, The Coppyeholdes Lyinge in Hynton are colored wth red, and the confines left whit. An^o: Dni 1549'. The writing on the map is indisputedly Christopher Saxton's, and although unsigned there can be no question of the attribution.

The size of the original map obviously cannot be given, but it is fairly large, the copy being approximately 85cm by 73 cm. Likewise colour cannot be distinguished, but it does not appear to have any large areas colourwashed. A scale rule, in the bottom left corner, is divided in Saxton's usual manner both horizontally and vertically, into eighty perches, marked numerically at 16, 22, 48, 64 and 80. On the copy the rule measures just over five inches which gives a scale of approximately sixteen perches to one inch, a scale favoured by Saxton. Above the rule is the statement 'This Scale contayneth 80 Perches.' and the whole is surmounted by open dividers reaching from 0 to 72 perches.

The map centres on a group of buildings including 'The Scite of the maner of Hynton', which, with the surrounding land, was occupied by Thomas Grimwood. His fields are named, together with the crops grown, i.e. 'Barlye', 'Wheate', 'Rye', and 'Otes', or simply pasture or meadow. In each case there follows a 'D', presumably describing demesne lands. Other occupants are named together with the word 'freholde', or 'cop' for copyhold tenants. The road marked 'The way from Dunwich to Hawleworthe', and 'the way from Dunwich to Bramfelde' runs across the north-east corner of the map, with 'hinton brooke' in the south-west, and 'spurling brooke' in the north-west. 'Hynton strete' in the south-east corner, with its four houses, borders on part of the manor of Westwood. There is a poignant comment beside one of the houses: 'Robart Brook tenament wch was Tho: Wades who hanged himselfe cop:'. There is no indication of any high ground. As land measurements are not given there may originally have been a written survey.

On the dissolution the manor of Hinton passed to the crown, and in 1538 was included in a grant of Blythburgh Priory to Sir Arthur Hopton, Lord of the Manor of Blythburgh, from whom it passed to his son, Sir Owen Hopton, Lieutenant of the Tower of London. In 1597 the manor was vested in Henry Gawdy and Henry Warner, probably as trustees for Sir Robert Brooke. At what date Sir Robert became possessed of the manor is not clear, but it is possible that Saxton was called in to map the manor on change of possession.

The manor, like that of Blythburgh, came into the hands of the Blois family about 1660 when the widow of the last of the Brookes married Sir William Blois, whose first wife was Martha, daughter of Sir Robert Brooke of Cockfield Hall, Yoxford.[25] Saxton's map has since remained with the Blois family at Cockfield Hall.

The map was probably made as a survey of the manor, no legal proceedings having been traced to suggest that the land was in dispute.

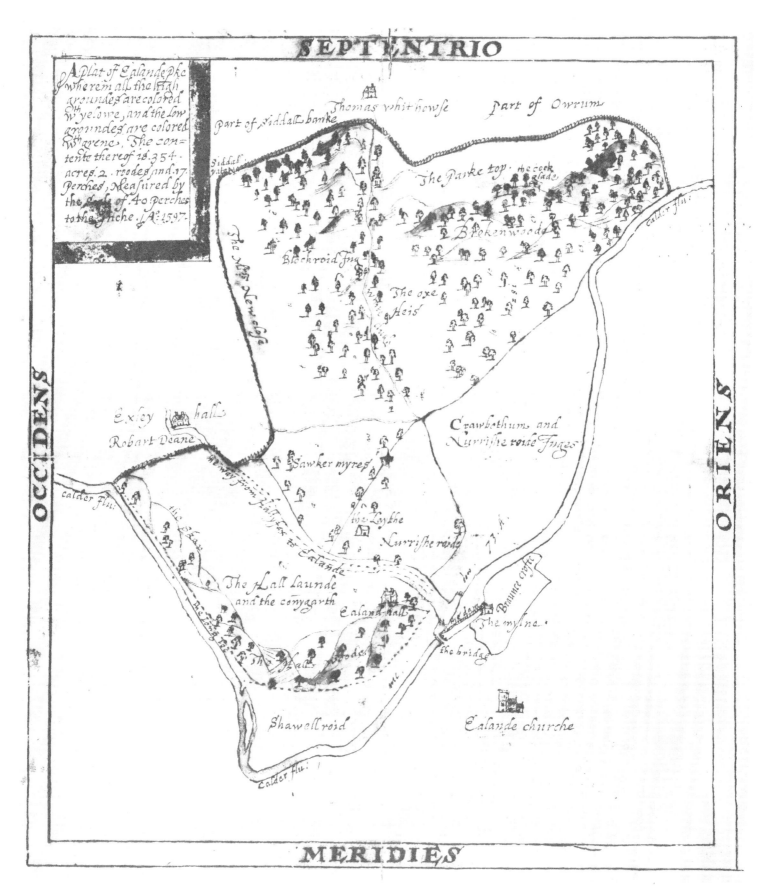

Manuscript map of Elland Park by Christopher Saxton, 1597.

Wadsworth; Elland; Hunsworth and East Bierley; Thornhill and other West Yorkshire estates, Yorkshire

In 1594 Christopher Saxton was employed by Sir George Savile on the first of many commissions for that family. It was to map the boundary between the manors of Wadsworth and Midgley, in the Pennine hills of the West Riding of Yorkshire, to the west of Halifax. Unfortunately the original map is lost and the only evidence of its existence is a tracing now in the Savile of Rufford collection at the Nottinghamshire Record Office.[26]

The tracing is one of three on the same sheet, two of which are of Saxton's maps, the third being a tracing from the Ordnance Survey of 1844. The tracings of Saxton's maps can undoubtedly be accepted as authentic from the portrayal of various features typical of Saxton's work, the writing which is recognisably his, and supportive documentary evidence. Both maps describe the boundary between the manors of Wadsworth and Midgley and are drawn as vertical strip maps.

The earlier map, numbered 3rd, has the title 'A plat of the bounders of the Maner of Waddesworth the same being colored with yelowe, The confines of Midgley wyth red, and the other confynes left whit, Anno Dñi 1594'. It is contained in a box cartouche with double inner lines to two sides, the top and right, and a single outer line; there are diagonal lines joining the outer and inner corners of the box as on the Hinton and Horbury maps. Written in the border at the bottom of the cartouche, is a rather flowery 'Christ: Saxton' followed by two flourishes of the pen. Beneath the title is a scale rule divided into 'halfe furlongs', 'Furlongs' and 'One myle', measuring four inches on the copy, giving a scale of four inches to one mile. Beneath the scale is written 'This Scale conteyneth one Myle'. There is no compass indicator and no border to the map.

The map itself shows the boundary between the manors as a line of short dashes, commencing at Oxnop (Oxenhope) Common in the north and running south to 'Fosterclugh heade', from whence it follows the watercourse downstream to its junction with the 'Calder flue' to the east of 'mythamroid brig', a distance of about three and threequarter miles. To the west of the boundary are drawn a series of Saxton's 'sugar-loaf' hills, perched on the summit of one of which is drawn a typical boxlike house. A series of turf pits are roughly drawn, including 'The turf pits in question'. In Foster Clough is a note 'Here haith bene a pinfould in the deuision betwixt Midgley and Wadesworth'. Wadsworth Common is shown to the west and Midgley and Warley Commons to the east.

The second tracing, numberd 2nd, is headed 'A plat of Wadsworth Common Colored with Yelowe the new Intackes are Colored with grene, the old Inclosures with red, and the Confines left whit, made by Christofer Saxton Anº: Dni: 1602'. There is no title cartouche. A simple scale rule, four inches long, shows a scale of four furlongs with the statement 'This Scale conteineth halfe a mile' giving an actual scale of one furlong to an inch, double that on the Wadsworth map of 1594. It is an extremely simple map, drawn down the middle of the sheet, and shows merely the dashed boundary line punctuated at intervals by named boundary stones drawn as vertically elongated triangles. Some are named as stones, e.g. Highbrownestone and Grenewood stone; others as mere stones (an alternative word for boundary stone), as in 'Ferrer merestone', 'Reasbye meare' and 'The merestone at Foster Clughheade'. The same area is shown as on the previous map, but on a larger scale covers a greater area on the paper. None of the decorative features, hills, turf pits, houses or hills are shown.

Accompanying the tracings are some explanatory papers. In 1844 it seems the exact position of the northernmost section of the boundary was again in dispute and a meeting of the freeholders of both townships was held at Hebden Bridge. Claims from both sides were submitted to Edward Durnford, Captain Royal Engineers of the Ordnance Survey, who recorded the minutes of evidence which were submitted to Thomas Lee of Wakefield, local solicitor to Lord Scarbrough, then owner of the manor of Wadsworth. Mr Lee was asked to produce evidence to support the claim of Wadsworth and he stated that the plans referred to '... had been written for to Mr Powell (Lord Scarbrough's Solicitor in London) but they had not been sent.' Durnford then produced some tracings furnished to him by Mr Powell, from the plans of Wadsworth 'made in 1594 and 1602'. In a letter dated June 22nd, 1844, Durnford wrote to Lee that he was sending a tracing for Lord Scarbrough on which was represented the boundary in question from three maps, and he describes the maps as those detailed above. It would appear then that the tracings now at Nottingham must be those made in 1844, and that the originals were then in existence. No trace of them can now be found. It is firstly implied that the maps were in London, held by Mr Powell; but Thomas Lee writing to Lord Scarbrough in July 1844 says: '... the two Plans made in 1602 and 1594 found by Mr Powell in the Evidence Room at Rufford and of which he sent copies to Capt. Durnford ...' which indicates the last known location of these two lost maps. The contents of Rufford Abbey were dispersed many years ago, but the Saxton maps were not included in the sale, nor are they in the possession of the present Lord Scarbrough.

At the meeting at Hebden Bridge a deed dated January 17th, 1596, was produced which concerned the division of lands and settlement of the boundaries in question. In consideration of the sum of £750 paid to John Lacy Esq., Lord of the Manor of Midgley, all suits, stryfes and variances between him and Sir George Savile, Lord of the Manor of Wadsworth and the other freeholders of Wadsworth were to be stayed and the boundary between the two manors was to be as therein described. In January 1598 Sir George Savile granted to the freeholders of Wadsworth the right of commonage, etc. 'on the West and Southwest side of the foresaid meares places and bounds', for a consideration of £40.

It would appear therefore, that Saxton was employed by Sir George Savile to draw his first map in 1594 when the question of exact location of the boundary was first disputed; and that he then drew another map in 1602 to describe the settled line.

The second commission executed for the Savile family was the mapping of Elland Park, south of Halifax and some twelve miles west of Dunningley. The map is in the Savile of Rufford Collection at Nottinghamshire Record Office.[27] It is unsigned but is endorsed 'Eland Parke Platt By Saxton 1597' in an unknown hand. The map is unquestionably by Saxton and is typical of his work. It is on paper and measures 31½cm by 26cm. The title, in the top left corner, reads: 'A Plat of Ealande Pke wherein all the high groundes are colored wth yelowe, and the Low groundes are colored wth grene, The contente thereof is 354 acres. 2 roodes and.17.perches, Measured by the Scale of .40.perches to the Inche. A^o: 1597.' The original colouring has faded, but the water course still retains its original blue. There is no scale rule or dividers, the scale having been explained in the title. (See Plate 13.)

The area drawn is contained in a rough V formed by the river Calder at this point, and is bound on the northern side by the park paling. The only land outside these boundaries is one small enclosure called Braunce crofte, to the east of the dam below Elland bridge, wherein was built 'The mylne'. The only road, shown as double dashed lines, is 'the way from Hallyfax to Ealande' which runs north-west from Elland bridge through the park to Exley Hall, then occupied by Robart Deane. 'Ealand hall', looking remarkably similar to Exley Hall, is situated south of this road about forty perches from the bridge; the only other building within the park is 'the Laythe', a barn. There is a range of hills in the north of the park, named 'Brokenwoode' at 'The Parke top', and a further range of hills following the course of the river in the southern extremity of the park called 'The Hall woode'. Trees are shown in profusion throughout the park. The park paling, which bounds the whole of the park not bordered by the river, is broken only by the road to Halifax at 'Siddall yate', drawn as the letter N with a crossbar between the first two strokes. Outside the park, Thomas Whit's house is shown at 'Siddall banke', and the church at Elland.

The map of Elland Park was presumably made as an estate survey, but whether it was for George Savile or Edward Savile is not clear. The Savile family was one of the wealthiest and most influential families in the West Riding of Yorkshire, originating in the south of the Riding. Sir John Savile of Tankersley married Isabella, daughter and coheir of Sir Thomas Elland in the fourteenth century, and through this marriage acquired the manor of Elland. The second son of this marriage, Henry, married Elizabeth, daughter and heir of Simon Thornhill, when the Thornhill estates came into the possession of the Saviles, and Thornhill Hall became the chief residence until the Civil War, after which Rufford Abbey became their principal seat. At the end of the sixteenth century, when Saxton mapped Elland Park, Edward Savile, a man reputedly of weak intellect, held the estates at Thornhill, Elland and elsewhere. According to Hunter,[28] Edward Savile had placed himself under the protection of George Lord Talbot, 6th Earl of Shrewsbury, who managed the Savile estates on his behalf. On the death of the childless Edward Savile in 1604, the main branch of the Saviles became extinct and the Thornhill, Elland, Hunsworth and other extensive estates were inherited by Sir George Savile of Lupset, near Wakefield, as nearest relative, complying with the terms of a settlement made in 1560 between George Talbot, Edward Savile of Thornhill and Henry Savile of Lupset, father of George Savile who was to marry Mary Talbot.

It may therefore have been through the recommendation of the Talbot family that Edward Savile employed Saxton, a suggestion more likely when it is learnt that Edward Talbot employed Saxton at about the same time. On the other hand, Saxton had already been employed by Sir George Savile when he mapped the boundary at Wadsworth.

Two years later, in 1599, Saxton was again employed by the Saviles when he drew a map of Hunsworth and East Bierley, between Bradford and Dunningley in the West Riding of Yorkshire, some six miles from Saxton's home. The map is also in the Savile of Rufford Collection at Nottingham.[29]

The map measures 40cm by 55½cm. The title, in the top right corner, reads: 'A plat of Hunsworth and Eastbyerle Common which is colored wth Red, the Inclosed groundes of Eastbierle and Hunsworthe adioyning are Colored with yelowe, and the confines Left white, made by Christ: Saxton An^o: D$\bar{m}$: 1599.' The scale rule, measuring two and a half inches, is drawn below centre on the left side, and is divided both horizontally and vertically into 100 perches, being marked at 40, 80 and 100. Above is written 'this Scale conteineth 100 pches', giving a scale of forty perches to one inch. The whole is surmounted by a very simply drawn pair of open dividers. The map is in good condition, the only noteworthy deterioration being where the ink, when used thickly in writing the cardinal points, has eaten into the paper.

The area shown on the map is basically adjoining the Wakefield to Bradford road, with parts of Tong, North Bierley, East Bierley, Hunsworth, Birkenshaw and Drighlington, but centring on Hunsworth and East Bierley Common, which was 'Inclosed and sowen'. The majority of the houses are drawn on the north side of the Wakefield to Bradford Road, and either the houses or their occupants are named. Tong Hall (which was rebuilt in 1702) and the church are shown in the north-east, but basically the map concerns the area from Tong Lane End to Dudley Hill. Coalpits are indicated on 'Eastbierle more', and one hill only shown, 'Wiscard hill', now known as Westgate Hill. Merestones (boundary stones) are shown as vertically elongated triangles along the boundary between Tong and East

Bierley from Westgate Hill to the boundary with Gomersal. No trees are shown. The colour has faded and is barely distinguishable.

As indicated in the title and the content of the map, it was obviously made in connection with enclosure of the commons. From west to east they are described: 'Tong Common Latelye Inclosed', 'Hunsworth Comon not enclosed', 'Hunsworth and EstBierle Common Inclosed and Sowen', 'pt of Tong new Inclosures' – the word new is rather squeezed in here – and 'Pt of Tong Inclosures'. To the north of the road is 'Part of Tong Inclosures'.

An inquisition taken after the death of Sir Henry Savile of Thornhill, who died in 1558 and who was Edward's father, states that he died possessed of Hunsworth Manor, held of Henry Tempest of Tong, and also 300 acres of land in Tong.[30] At the Nottinghamshire Record Office is a written survey, not attributed to Saxton, entitled 'Survey of Hunsworth, Southoram & Newe hall 1604', which lists tenants of the Saviles' and their holdings.[31] Several of the tenants can be identified as the same as named on Saxton's map of 1599. A note of interest on the 1604 survey concerns measurements employed by the unnamed surveyor: 'wood grounde' is given in acres, roods and perches appertaining to each tenant, and dayworks are not used. Below are notes declaring that a certain tenant's wood 'by the long measure conteyneth ...', with acreage given. In one case is written 'by the greate measure conteyncth ...' with acreage given. A quick mathematical comparison between the original measure, and those by the long or great measure, reveals that the former are approximately 1·6 times greater than the latter.

There is one further map by Saxton in the Savile collection at Nottingham;[32] it is of Thornhill, the principal seat of the family, which lies south of Dewsbury in the West Riding of Yorkshire, some five and a half miles south-west of Dunningley. The map is unfortunately damaged and the greater portion of the title missing, including Saxton's name and the date. It is, however, unmistakably by Saxton, and three copies of a written survey of Thornhill by Saxton, dated 1602, also survive, suggesting that the map was probably made at the same time.

The map is comparatively large, measuring 66½cm by 144cm with the bottom left corner completely missing. There is an additional section, 37½cm by 38½cm adjoining and above the top left portion of the main map. A large portion of this latter section is also missing, as is much of the top section of the sheet, and much of the right side. Fortunately the portions illustrating Thornhill Hall and the demesne lands are clearly shown. The cardinal point 'WEST' is missing and only the letter N remains of the word 'NORTH'. All that remains of the title is: 'A plat of the Dem ... ... des Belonging ... col ...', which could be expected to read 'A plat of the Demaine groundes Belonging to the Manor of Thornhill ... coloured ...'. It would probably also state the date and that it was made by Christofer Saxton, and what the colouring indicated. There is adequate space for about six lines of writing. There is no scale rule shown, this was probably in one of the areas now missing.

The area shown on the map centres on Thornhill Hall and stretches from Dewsbury Church in the north-west to Horbury Bridge in the south-east. The river Calder, with a total of five islands which Saxton calls Steaner, forms the northern boundary of the area mapped. One island, immediately west of Horbury Bridge is marked 'The Steaner in question', but this is not the same as that, also at Horbury, which Saxton was employed to map in 1589, which was a little distance down stream.

Thornhill Hall with its surrounding moat and adjacent orchards, washing garth, bowling place, dairy, stables and ox house, is surrounded on three sides by paling, being situated between the empaled Thornhill Park, with its solitary lodge, to the north, and the empaled New Park, with its two fish ponds, to the south. No representation of high ground is shown, but trees are drawn in profusion, particularly on field boundaries. All the fields are named, but no measurements given. (See illustration p. 77.)

Accompanying the map is a written survey by Christopher Saxton of which there are three known copies. Only one is ascribed to Saxton in the title, but is not in his hand; a second is in Saxton's hand but is not ascribed to him in the title, although attributed to him in an endorsement; the third survey is a copy, with later additions by Robert Saxton. For convenience they will be referred to as (1), (2) and (3).

(1) is a survey at the Savile Estate Office, Thornhill and is entitled 'A Survey of the Demaine Groundes belonging to the Mannour of Thornhill taken by Christopher Saxton Anno Dni. 1602'. Below the title, added in pencil, is written: 'A Survey of Several other Estates follows'. The whole is sewn together as a booklet of twelve sheets giving twenty-four sides of writing. It is not written by Saxton. Land measurements are given in acres, roods, dayworks and perches in columns to the right. The first two pages list the demesne groundes of the manor of Thornhill which are identical to the fields named on the map described above. On pages three to six are listed the closes of Thornhill outside the demesne grounds, together with the tenants' names. Indication is given whether the land is enclosed or in the open field. Some tenants have a number beside their name, possibly suggesting that a map with corresponding numbers had been drawn, or was intended to be drawn. The total acreage held by tenants in Thornhill was 325a or 1dw 2¾p.

On the following pages are found similar lists of tenants and their holdings in Middlestown, Emley Park, Emley, Skelmanthorpe, Elland, Stainland, Botham Hall, Rushworth Hall, Rushworth Turner and The Mill, Rushworth Mosselenden, Rushworth The High Mosse, Barkisland Goslingroide, Stansfield Rawtenstall, Waddesworth Widduppe and Wadsworth Walshey. Each page is headed with the name of the manor or estate described, but unfortunately on this copy most are partially or totally obliterated by the ink having rotted away the paper. However, most names can be established by reference to another similar survey of 1604, not attributed to Saxton, at the Nottinghamshire Record Office.[33]

Saxton does not indicate what scale measure he employed in his survey, but in that of 1604 mentioned above, in which the acreages listed are the same as given by Saxton, there is a statement that the statute measure was employed. One tenant in Skelmanthorpe apparently paid a yearly rent, but upon the survey 'could not find any ground'!

(2) The second survey of Thornhill is at the Nottinghamshire Record Office, and is written by Saxton on a single sheet of paper.[34] It is headed: 'A Survey of the Demaine groundes belonging to the Manner of Thornhill Anno Dni: 1602'. There are two endorsements, firstly in Saxton's hand: 'A Suruey of Thornhill Demaines', and secondly: 'The Survay of Thornhill Demaynes taken by Christopher Saxton Anno 1602' in an unknown hand. The content of this survey is identical to (1) described above in the section relating to Thornhill demesnes only, but with the addition of a final line giving the total content of Thornhill Demaines as 835a or 3dw 2p. Saxton appears to have omitted two closes from the survey which appear on the map, Hobroyd and Persroyd, with a combined acreage of 45a 3p 0dw 8p. None of the other estates are included in this copy of the survey. This survey is reproduced (Plate 14) to illustrate both Christopher Saxton's writing and a typical survey.

(3) The third survey of Thornhill is also at the Nottinghamshire Record Office.[35] It is headed: 'A Suruey of the Demaine groundes belonging to the Manner of Thornhill Anno Doñi: 1602.', with '1602 Survey of the Demesne lands of Thornhill' endorsed on the cover in a different hand. The survey is not written by Saxton, although the writing is similar. The content is identical to (1) and (2) above as far as Thornhill demesne lands are con-

A Survey of the Demaine groundes belonginge to the Manner of Thornhill. Anno Dni 1602	Acres	Rodes	Dayworkes	Perches
Inprimis the Scite of the Manner within the mote	0	3	6	0
The walkes about the mote	0	2	2	0
The wasshing garth	0	0	5	0
The New Orcharde	1	2	5	2
The fouldes	3	3	9	0
The Bowling place	0	2	2	0
The Olde Orcharde	2	2	6	2
	1	0	5	2
The new parte	3	2	7	2
Nichson close	24	0	6	2
The Flatt flatt	9	0	6	2
Mowschole close	6	1	6	2
Sunyeard Spring	5	1	6	2
Hullis closes	23	2	7	2
Ladye milne Ing	4	0	7	0
Long hastingleye	28	2	7	0
Nether Swarthingforde	24	1	9	0
Over Swarthingforde Ing wth the Steanarde	35	3	9	2
The milne closes on the South side the wayes	27	1	1	0
The milne closes on the north side the way	19	1	5	2
Hallowes Ing	4	2	2	0
The Oxe close	17	1	5	0
Bushge close	11	2	7	0
The launde	30	3	1	0
The Steanard	1	0	7	2
The long close	4	2	7	1
Pearetre flat	8	0	8	0
The broade oke close	15	2	8	0
The Leies	19	0	3	0
The Longleies	12	0	8	3
The Sandes meadowe	12	2	2	3
Thornhill parke	153	2	5	0
Stringer closes	7	3	2	2
The Sr Holme meadowe	9	2	3	0
Mille crosse	42	2	2	3
Tomson closes	9	3	8	2
The Headefeilde	213	2	5	0
William Hall Headefeilde close	2	2	8	2
The Steaner in Gestion next unto Horburye brig	8	0	7	1
The new Intak	5	0	8	1
part of the parsons close within the parke	2	2	0	1
The parsons laith close				
The content of Thornhill Demanes	838	0	3	2

Written Survey of Thornhill demesne by Christopher Saxton, 1602.

cerned, though omitting the final total acreage as found in (2), and with several additional lines. This survey includes the missing closes of Hobroyd and Persroyd and four others, two of which were 'now inclosed', according to a note on the back. This note, dated April 23rd, 1612, is signed 'by me Robert Saxton' which is interesting as being the only evidence that Robert Saxton was employed independently of his father by the Saviles.

Isle of Axholme, Lincolnshire and Nottinghamshire

In 1596 Christopher Saxton was a witness in connection with a boundary dispute between the crown tenants in Haxey and Owston in the Isle of Axholme, and the crown tenants in Misterton, part of the Duchy of Lancaster, which basically concerned the exact position of the boundary between Lincolnshire and Nottinghamshire. The Isle of Axholme lies east of Doncaster, almost adjoining Hatfield Chace, later surveyed by Robert Saxton, and some thirty-five miles from Dunningley.

The case was brought to the Court of the Exchequer and a Special Commission was appointed in May 1596; the enquiry was to be held at Stockwith 'to examyne witnesses on both partes ...' and states 'And that also a joynte plotte or s'vall plotts be taken and made p'pectlie by scale and measure as well of all the p'ties lande & comons of Haxey & Owston, & all the places knowen meeres meets divisions & boundaries about & within the same comons, as also of all the p'tes landes & comons of Stockwith & Misterton, and of ev'ie place and places of knowen name & diffiente in the said townshipps lands & comons & that the p'son or p'sons that shall tke, make, & sett forthe the said plotte or plottes shall be sworne by the said Commissioners ..., that he or they have taken, made and set forthe the said plotte or plotts trewlie & indifferente soe neare as by arte or skill he or they came or maye doe the same without affec'ion or fav'r of annie p'tie or p'ties of the said townshipps or annie of them.' It was also stipulated that fourteen days' notice be given before 'the takings of the said plotte or plotts & sittinge of the said Commission ...'.[36] Clearly evidence portrayed on maps in cases such as this was of prime importance.

Depositions of witnesses were taken at Stockwith on September 18th, 1596, and amongst those called were Christopher Saxton and a certain Richard Heley of Crosbye in Lincolnshire, described as a gentleman aged thirty-two. Heley and Saxton both exhibited maps to the Commissioners, but only one, that by Saxton, survives.[37] Unfortunately the title and attribution on the map are missing, but an examination of the style and writing leaves no doubt as to the author.

Saxton's deposition was as follows: 'Christopher Saxton of Dunningley in the pishe of West ardesleye in the countye of Yorke gent of the age of ffiftye twoo yeares or thereabouts sworne & examyned the day & year above said deposeth and sayeth That the plott now by him exhibited to the Commissioners and herwth certefied ys according to his skill and knowledge and the informacion of the meetes lymittes and Bownders of the Comons and wast growndes in variance betwixte the ptyes to this suite and also of others mencioned in the order annexed to the Comission ys trew wthout any ffavour or affecion to any of the said ptyes.'[38]

The map itself is $33\frac{1}{4}$cm by $47\frac{1}{2}$cm, although much of the top, including the title, is missing. Presuming the cardinal points to have been centrally positioned in the margin, as was usual, the map must have measured about $55\frac{1}{2}$cm vertically when complete. To the bottom right is the scale, a divided rule two inches long, marked at 2, 4, 6 and 8, followed by the word 'furlonges'. Above the rule is written 'This Scale conteineth one Myle', giving a scale of half a mile to an inch. The area mapped is bound to the west by the River Idle in the north and the River Eae in the south, by the River Trent in the east, with parts of Gringley Carr and Walkringham in the south, and with Haxey North Carr and Owston Common in the north. The area is traversed by the watercourses Bickerdike River and Burnham Skires which runs into The Trough and Lounds Sound. The villages of Newbie, Haxey, Owston, Mysterton and Kelfeilde are shown. Several windmills, river bridges, some trees and an occasional named hill are drawn. The position of the disputed county boundary runs parallel to, but north of the River Bickerdike. It is a very detailed and interesting map, naming all natural features, closes, enclosures and commons. The watercourses are painted blue and the commons colourwashed in brownish-yellow. An exact copy of the map is reproduced by Dunston in *The Rivers of Axholme*, together with a history of the continued disputes in the area.

Manchester, Lancashire

In 1596 Christopher Saxton carried out a survey of Manchester, which had it survived, would have been the first of only two maps of towns by Saxton known to us. There is no modern record of this map having been seen, and the only record of the work is of contemporary date, in a diary kept by 'Dr' John Dee who was installed as Warden of Manchester in February 1596. John Dee (1527–1608) was a mathematician and astrologer who took a great interest in cartographical development, visiting scientific centres in Europe where he became a friend of Mercator, and disciple of Phrysius (see page 43).

In his diary Dee recorded personal details, employment of staff and wages paid, etc.[39] On June 14th, 1596, Dee was visited by Mr Harry Savill, the antiquary; this would be Henry Savile of Shaw-hill, Skircoat, Halifax, commonly known as Long Harry Savile, an eminent scholar and friend of Camden. The relevant entries for the period are as follows:

June 15th 1596, I wrote by Mr Harry Savill of the book dwelling at Hallyfax to Christopher Saxton at Denningley.

June 21st, Mr Christopher Saxton cam to me

July 5th, Mr Savill and Mr Saxton cam

July 6th, I, Mr Saxton and Arthur Rouland, John and Richard to Howgh Hall

July 10th, Manchester town described and measured by Mr Christopher Saxton

July 14th, Mr Saxton rode away.

It is not clear from Dee's entries whether Saxton drew a map, wrote a survey, or indeed did both, but several points of interest can be deduced from the entries. Firstly that Saxton was living at his home at Dunningley at this time, and that he was available to answer Dee's summons quickly, arriving within five days of the letter having been written, presumably less from its receipt. Manchester is about thirty miles from Dunningley as the crow flies but over the rough terrain of the Pennines the distance travelled would be considerably more. How long Saxton stayed in Manchester on his first visit is not recorded, presumably long enough to ascertain the nature of the proposed survey and settle terms. His second visit lasted nine days, but from the entry Dee made on July 10th it may be that the work was finalised that day by Saxton drawing up his map. Alternatively it could be interpreted that Saxton physically worked at his survey on that day only; it is an ambiguous statement, but it is highly unlikely that Saxton could have completed a survey of Manchester town in one day only when it is clear that a perambulation of the parish boundary the following year took six days. It depends really on how large an area was surveyed. The final point of note is that Saxton obviously stayed with John Dee in Manchester. It is clear that he was an acceptable guest at the houses of his employers; in this case Saxton and Dee would have many mutual acquaintances and would have had much in common to discuss.

Ackworth, Yorkshire

In the same month, July 1596, Saxton surveyed the manor of Ackworth, three miles south-west of Pontefract in the West Riding of Yorkshire, some eleven miles south-east of Dunningley.

The written survey, now in New York, is headed: 'A Suruey of the demayne Landes of the Mañer of Acworth made (here 'made' is crossed out and 'taken' written above) by Christofer Saxton In July. Anº: Dni 1596.'[40] It is not written by Saxton.

The survey names fields and gives acreages, but no occupants. It is divided into three sections describing the pasture land, of roughly 89 acres, the meadow, of roughly 41 acres, arable land of 26 acres and the field land of just over 25 acres, giving a total of 182a 1r 8dw 1p surveyed, according to Saxton, but which in checking the additions is found to be three acres too great for the meadow land. There is a note at the end that '... the common pasture wherein the lord haith xl yaites is not surveied nor the milne neither the common' and signed 'Christ: Saxton'.

Ackworth was in the Honour of Pontefract and part of the Duchy of Lancaster, adjoining Tanshelf to the north, later surveyed by Robert Saxton. In 1567 the manor of Ackworth was leased by James Corker for a term of thirty-one years. Corker died in 1573 and beaqueathed his 'leases and termes of years' which he had in the hall and demesnes and mill of Ackworth to four of his sons. Martin Frobisher was the next leaseholder of the demesne lands at a rental of £10 10s 8d. At what date Frobisher took over is not known, but it may have been the date at which Saxton was employed to survey the manor.

The manuscript was formerly in the possession of Sir Thomas Phillips and was sold by Sothebys in 1903, when Maggs Brothers of London purchased this and the survey of Sandal by Robert Saxton for £1 11s 0d.

Whitchurch, Shropshire; Shafton, Yorkshire

In March 1596–7 Christopher Saxton and Henry Caldecott were together engaged to survey lands in Shropshire for Edward Talbot. The written survey, at the Salop Record Office,[41] is headed: 'The particular Survaye of the demeayne lands and tennements Beloning unto the lordshipes and mannors of Whitchurche Dudingtō Linnyall and Marberie taken in Aprill Añº. dñ 1593. in the time of the honorable Henry Talbott Esquire And augmented by the travell of Christofer Saxton and Henry Caldecott in the moneth of March Año dñi 1596 & 1597 in the xxxix[th] yere of her Maties most happie Raigne And Nowe beinge pcelles of the Possessiones of the honorable Edward Talbott Esquire: as hereafter followeth.' The survey is not written by Saxton, nor is the copy which accompanies it.

Townships are given, with the names of the tenants and sometimes, but not always, the names of the fields occupied, together with land measurements in acres, roods, dayworks and polls (the same as perches). The value of each tenant's holding and the total acreage and value have been added in a different hand.

It is rare to find a record of Saxton working with another surveyor. Henry Caldecott surveyed Hose in Leicestershire, 1603–6, but nothing else is known of his work. How much of this survey was original work cannot be ascertained as the survey of 1593 mentioned in the title cannot be traced.

Henry and Edward Talbot were sons of George Talbot, 6th Earl of Shrewsbury, and presumably Henry Talbot inherited these Shropshire estates on the death of his father in 1590. In 1596 the lands had been vested, by trustees, in Edward and in 1598 he sold them to Sir Thomas Egerton.[42] The revision of the survey must have been made prior to the sale to Egerton.

In 1597 Christopher Saxton was again employed by Edward Talbot, this time to map the Lordship of Shafton in the West Riding of Yorkshire, to the south of Cold Hiendley, which he was later to survey for Michael Wentworth, and thirteen miles from Dunningley.

This is one of the largest maps, being 123½cm by 94cm and was drawn on four sheets of paper. The title, below centre to the left, reads: 'A Plat of the Lordshipp of Shafton Wherein the Inclosed groundes are colored wth grene, The feilde groundes are colored wth yelowe, The Comons and waste groundes are colored wth Red, The Inclosed Freholdes, and confines, are Left Whyte. Made by Christofer Saxton. An^o: 1597.' Below the title is the scale rule measuring seven inches, divided horizontally and vertically into 112 perches, being marked numerically every eight perches. The first forty-eight perch section only is surmounted by open compass dividers, between the tips of which is written: 'A Scale of Perches', giving an actual scale of sixteen perches to one inch. The map is in very good condition, the writing clear and entirely legible.

The land shown is bounded to the north-west and north by an unnamed watercourse crossed in the west by 'Kirkbrig' and in the north by 'Sandall bridge' (now corrupted to Sandy Bridge). The whole of Shafton township is shown, with adjoining townships named. Field names are given, together, in most cases, with the tenant's name. No acreages are given. There is no woodland as such, but trees are shown on the field boundaries. There is no indication of high ground. The watercourse still retains its blue colouring, and some yellow is still distinguishable, but the rest of the colour has faded.

With one exception all the buildings are within a quarter of a mile of Shafton Hall which lies in an orchard on the west of the road through the village. The houses fronting the village streets all have long strips of land behind; on the north is 'The Grene' with its enclosed pinfold. The five town fields are named: Lydyate Feilde, Eshbarrowe Feilde, Towne ende Feilde, which surrounded the village, and The Nether Feilde als Colepit Feilde, with three 'colepitts', and The Heade Feilde to the south. The pits are drawn as large dots and are interesting in that mining is not otherwise recorded here until later years, although Shafton is now completely absorbed as a mining village. Enclosures had already nibbled into the open fields. Enclosed and unenclosed roads are distinguished by continuous or dotted lines.

In 1580 George Talbot, 6th Earl of Shrewsbury, purchased the Lordship of Shafton, together with a considerable quantity of other land in the area. The Earl, custodian of Mary Queen of Scots, had been encouraged by Queen Elizabeth to purchase land to demonstrate that he was not impoverished by his 'entertainment of the Queen and her retinue'. Shafton was settled on the Earl's youngest son Edward Talbot, who must have employed Saxton to draw the map as a general survey of his holdings. It seems quite likely, in view of the fact that no acreages are given, that a written survey may have been compiled to accompany the map, but if such were made, it has not survived. Edward Talbot died childless in 1617 and his estates passed to the descendants of his sister Mary (Savile), and through them, at the end of the eighteenth century, to the Foljambes of Osberton, in whose possession the map remains. There is a copy at the Nottinghamshire Record Office.

Horbury and Netherton, Yorkshire

In 1598 Saxton was again called upon to draw a map of land in dispute. The case arose between the copyholders of Horbury and the freeholders of Netherton whose townships

PLATE 15

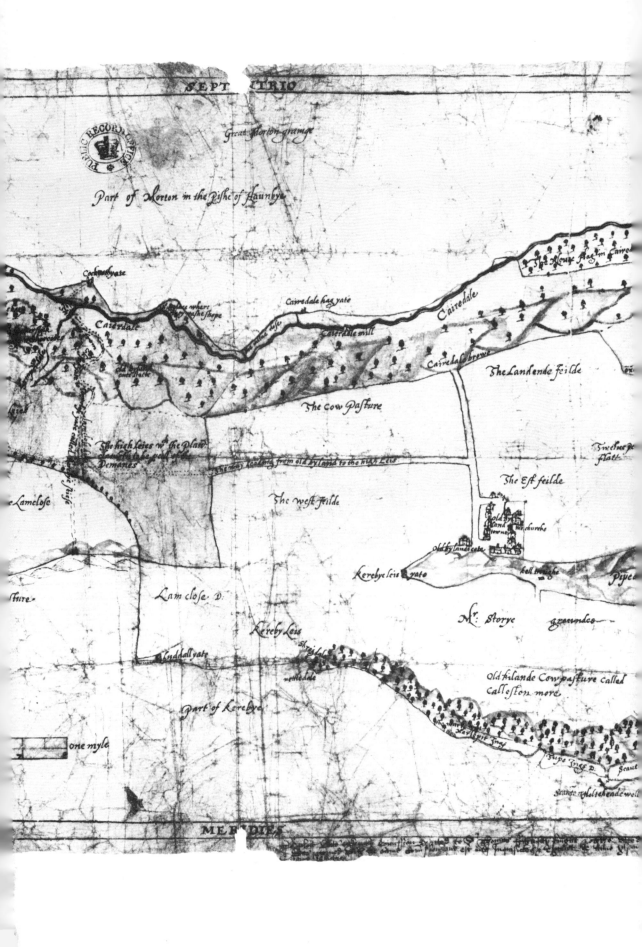

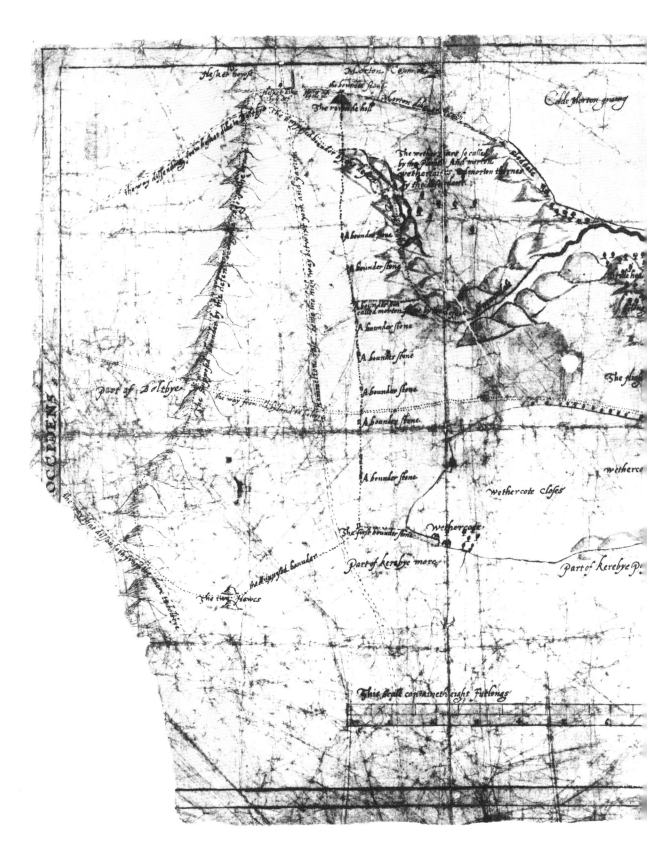

Manuscript map of Old Byland by Christopher Saxton, 1598.

were divided by the River Calder, west of Wakefield in the West Riding of Yorkshire, and only six miles south of Dunningley.

Saxton's map, now at the Public Record Office, measures only 27½cm by 28cm, and is the smallest of Saxton's maps.[43] It simply shows the course of the River Calder, its diverted course and an island called the Steaner which had been formed as a result of diversion of the river. The title, at the bottom left of the map, reads: 'A plat of the steaner in controuersye betwixt Horburye and Nethertō being colored wthgrene and the confines left white Made by Christofer Saxton Anº: Dni. 1598.' A scale rule to the right measures two and a half inches, being divided into perches and numbered at 16, 32 and 40, giving a scale of sixteen perches to an inch. Above is written 'This Scale conteineth 40 pches'.

The river Calder is shown across the map, with Adingforth pasture in Horbury to the north, and the Stend pasture and Brode Ing, both in Netherton, to the south. The old course of the river flowed in a gentle curve, but a new course had been carved out to the north, forming a U bend and isolating the steanor of 7a 1r 8dw according to Saxton. A weir is shown on the map at the entrance to the new course. All the original colour has faded and there are no cartographic details of buildings, trees, roads or high ground.

Several documents, late in the possession of the Horbury Common Lands Trustees, tell the legal story of the dispute. The copyholders of Horbury, part of the Manor of Wakefield and of the Duchy of Lancaster, had for time immemorial had common pasturage in Horbury which was separated from Netherton, or Nethershitlington, part of the Honour of Pontefract but also in the Duchy of Lancaster, by 'an anncyent Ryver called Calder the anncyent and mayne streame whereof had allways runne in one currante or Course until of late within few years that ptelye by vyolence of the water and ptlye by loosenes and softnes of the groundes and banckes towards Horburye syde some little pte of the streams had broken throughe the soyle of the sayd Comon of Horburye syde and lytle by lytle had made a small newe currante through the same inclosinge by that meanes pte of the soyle of the said Comon of Horburye to the quantetye of ffowerteen acres or thereabouts beteene the sayd newe made Currante and the sayd anncyent streame of Calder.'

The Horbury men claimed that it was dangerous to take their cattle to the island so formed and rebuilt an 'ould decayed weare' at the mouth of the new stream in an effort to redirect the course of the river. The Netherton men reacted by attempting to destroy the weir, whereupon the Horbury men took recourse to law. The case went before the Manor Court at Wakefield, the Court of Common Pleas at Westminster, the Assizes at York and the Duchy Court at Westminster. The legal wranglings went on for several years with Horbury claiming that the Steaner was their land by custom and Netherton claiming that the townships were divided by the river and that the Steanor land was now theirs. Finally judgement was given in the Duchy Court for the Horbury copyholders.

The three documents examined were all signed by Gerrard, who was William Gerrard, Clerk of the Duchy. Saxton's map shows the Steanor to be just over seven acres, whereas the plaintiffs claimed it to be about fourteen acres. Even allowing for different basic measurements, the discrepancy is considerable; possibly the Horbury men were trying to strengthen their case by exaggeration, or perhaps were simply inexperienced in judging land measurement.

It was probably as a result of this and other disputes that, in 1653, the Horbury Common Lands Trust was formed, a body which still flourishes.[44]

Old Byland, Yorkshire
In July 1598 Christopher Saxton was once more called upon to draw a map in connection with disputed boundaries. The request came from the Court of Exchequer and the land in

question was in Old Byland, on the North Yorkshire Moors. The map, now at the Public Record Office, is one of the most detailed of Saxton's maps and covers an area roughly four and a half miles from west to east, and two miles from north to south.[45] (See plate 15.)

The map measures 41cm by 92cm with the two bottom corners missing, but nothing of significance is lost. The title, abutting the top right corner, reads: 'A plat of the pishe of Old Bylande wherein all the Inclosed groūdes are colored wth red, The comōns are colored wyth yelowe, The Defendants bounder betwixt Oldbyland and Morton, is drawen with a grene lyne; and the confines left white. Made by Christofer Saxton. An[o]: Dm:.1598.' with 'p Sacramer' (per sacramentum – on oath) written below. Beneath the title is a note: 'that Christofer Saxton as well by the consent and apointm[t] of the Comission[rs], as of the Solicitors of the pties p[t]: and def[ts]: did make this Platt, and was sworne that the same was trulie made to the best of his skill.', and this was endorsed by the commissioners. A very simple scale rule to the bottom left, measures eight inches and is marked at half furlongs and numbered from one to eight, being followed by the words 'one myle', giving a scale of eight inches to one mile. Above is written 'This Scale containeth eight Furlongs'.

The area drawn is bounded by Caydale and Murton to the north, Boltby to the west, Kirby and Scawton beck to the south, and the river Rye and Rievaulx to the east. Trees and woods are shown in profusion; buildings, watercourses, roads and tracks are all detailed. On 'Cairedale browe', to the north of Old Byland, the hills have been drawn upside-down which gives a rather unusual appearance. Boundary stones are drawn as upright rectangles. 'Old byland towne' with its church and 'cote' is clearly drawn in the south-west corner of the 'Est feilde'. This was in fact not the original position of the village; in the north-west, at the junction of the Caydale and Ryedale, Saxton shows a 'Tylehowse', and it was here that Byland village was situated in the eleventh century. Monks settled at the head of Pipedale just south of the present village in 1143, Beresford suggests at the point marked 'kell and trough' by Saxton.[46] However, Rievaulx Abbey was only two miles distant and the bells of each house could be heard by the other; seeking more fitting solitude the monks of Byland moved, finally settling at the site where the ruins of Byland Abbey can be seen today between Ampleforth and Kilburn.

It was as a result of the dispersal of the Abbey lands that conflict arose. Old Byland had passed to two families; in 1598 Edward Wotton owned most of the parish of Old Byland, and Sir William Bellassis owned the neighbouring parish of Murton and an estate called Wethercotes in Old Byland. The disputed land was called Wetherlayers which each claimed as their own. A law suit was commenced and in July 1598 an order was issued by the Court of Exchequer for 'a perfect plott of the places and groundes mentioned in the pleadings' to be made. Depositions were taken at Byland in late August, and Saxton must have been present for him to have gained so much information which he then translated onto his map. The evidence was in favour of Sir Edward Wotton, but in fact, although the arrival of the depositions and the map in London are recorded in June 1599, there is no judgement recorded.

The lands disputed were in the west of the parish and Saxton shows each boundary claimed by the parties, together with all the boundary stones mentioned in the depositions, two portions of Hesketh dike claimed by each side, supposed bounders and different landmarks.

Maurice Beresford has written in great detail about the map and the case in *A Journey Along Boundaries*, from which much of the above information has been taken.[46]

Luddenden, Yorkshire

Two maps of Luddenden Foot by Christopher Saxton survive in the Public Record Office, both made in connection with legal disputes over water rights.[47] Luddenden lies in the Pennines, four miles west of Halifax and seventeen west of Dunningley as the crow flies.

The earlier map measures 51cm by 36cm, with the title above centre on the left. It reads: 'A plat of the course of Ludingdē brooke wth the milne goites taken forth of the same from Luddingden Chappell to Luddingden foote, wherein the Confines of Midgeley is colored with yelowe, and the confines of wareley is colored wth red, made by Christofer Saxton. Anno Dñi 1599.' The line scale, to the bottom right, measures four inches, is marked in perches and numbered at 16, 32, 48 and 64, giving a scale of sixteen perches to one inch. Above is written: 'This Scale containeth 64. perches', and the whole is surmounted by open compass dividers.

The map shows the unnamed river Calder in the south-west, with Boy Brig and Boy milne downstream. 'Luddinden brooke' is traced from 'Luddinden chappell', a handsome looking building in the north, to its confluence with the Calder in the south, dividing the townships of Midgeley and Warley. Several goits and dams are shown throughout the whole length of just under a mile. Some of the fields are named, and occupants of buildings and owners of land bordering the brook are given. No trees, woods or hills are shown. Most of the colouring has faded, with the exception of the watercourses and buildings. At 'Luddinden foote', on the Warley side, is a large house, much larger than any other, named 'Michaell Foxcroft howse' (in fact it was Michael Foxcroft junior who lived there), and above is 'the way frō Brearley to Halifax'. North of the road are 'Michaell Foxcrofte fulling milnes' on his own goit. On the other side of the brook, just in Midgeley township, stands 'Mr Farror fulling mylls' on his goit. A little to the north is 'Mr Farrar dam' and 'The dam and breach in question'. Foxcroft's dam is situated yet further upstream and the Queen's mill and the site of yet another old mill and goit are shown beyond.

The second map, made in 1601, is unattributed, but undoubtedly by Saxton bearing all his characteristics. It measures 44cm by 29cm. The title is placed to the top left and reads: 'A plat of Mychaell Foxcrofte groundes with his howse Milnes, goite or dames Anno Domini 1601.' To the bottom right is the scale rule measuring three inches, divided horizontally and vertically into perches and numbered at 16, 32 and 48, with the statement 'This Scale conteineth 48 pearches' above, the whole being surmounted by open dividers. The actual scale therefore being sixteen perches to one inch, the same as the earlier map.

The content of the 1601 map however, differs slightly in only showing the area from the Calder to the head of Foxcroft's goit, about half the distance upstream shown on the earlier map. Foxcroft's house is named 'luddingdeine roide howse, als Luddingdiene foote howse', and his mill is also shown. Surrounding land is named, but only one other landowner is identified. Farrar's mill is drawn, but not named. Trees and woodland are shown on this map, and the road to Halifax as on the 1599 map.

The full course of legal proceedings centring on Luddenden Foot can be followed in the papers from the Duchy Court at the Public Record Office.[48] There existed a group of wealthy landowners who appear to have spent their lives in continual conflict; there were numerous struggles in and out of court in which large sums of money were expended, and on several occasions blood was shed and life brutally taken. The main antagonists which concern us, were Henry Farrar, a Justice of the Peace, and Michael Foxcroft, both of whom had fulling mills at Luddenden Foot at the close of the sixteenth century. Michael Foxcroft and his son Michael had been involved in a treacherous wrangle over land in Warley with the former's brother-in-law Samuel Wade, nephew of Henry Farrar, which ended with Foxcroft killing Wade, but not settling the question of the disputed land.

Henry Farrar had two fulling mills on the Midgley side of Luddenden brook and had constructed a dam to serve them. Michael Foxcroft had his mill nearby on a goit which took water from the brook higher upstream, so diverting water from Farrar's mill and preventing it from working. There were lengthy court enquiries and proceedings and Christopher Saxton was called in to map the whole area in 1599. According to Foxcroft, Farrar's men had twice forcibly broken down his dam and filled the goit, the second time leaving an armed guard for three weeks to prevent Foxcroft effecting repairs. Foxcroft claimed to have lost 100 and 3 score kerseys each week by stoppage which would have been brought to be fulled. Farrar claimed, by grant of Letters Patent, that he could take sufficient water from the brook to drive his mill and his servants had turned the water into its ancient course. The Duchy Surveyor for the Northern Parts, Edward Stanhope, heard witnesses and ordered Foxcroft to reimburse £20 for each year Farrar's mill had been stopped, which he failed to do. Two further maps were submitted to the Duchy Court; one in 1601, the unsigned one by Christopher Saxton, and another in 1602 by John Bell of York.

Stanhope records taking John Bell 'a man expte both in surveigh and takinge levell for water' with him, but no mention is made of Saxton in the surviving documents.

The story is told in greater detail by Hugh P. Kendall in *The Story of a Local Feud*, from which much of the above was taken.[49]

Woolley; Notton; Cold Hiendley, Yorkshire

In 1599 Christopher Saxton was engaged by Michael Wentworth, who in that same year purchased estates in Woolley and Notton, six miles south of Wakefield in the West Riding of Yorkshire, and some nine miles from Dunningley. A written survey of Woolley and a map of Notton survive, both being dated 1599. There is also a written survey of Cold Hiendley, to the east of Notton, made for Wentworth, probably in 1605.

The survey of Woolley, now at the Brotherton Library, Leeds University,[50] is headed: 'A Suruay of the (illegible) of Woolay made by Christofer Saxton Aº 1599.' The fifth word is illegible, but must have had about six letters, the first of which might be an S. The survey is written on three and a half sheets of paper and is in rather poor condition, being badly creased, but nevertheless generally legible. It is written in an unknown hand, with the exception of the title and four lines which could have been penned by Saxton. There are additional notes in yet another hand at the end concerning land purchased since the survey was made. Measurements are given in acres, roods, dayworks and perches.

The survey details the demesne land of Woolley, which total 397a 2r 7dw 1p, and then lists twenty tenants and their holdings, which total 784a 2r 3dw 1p. Most of the tenants had land in the common fields – the Kirke feilde, the Netherfeilde, the Equall feilde, the Church feilde and the Lyttle feilde – and their other lands are described as being meadow, arable or pasture land. The numerical additions are fairly accurate, but not completely so.

The map of Notton, at present on deposit at Wakefield City Museum, measures 115cm by 147cm, and is drawn on paper now mounted on cloth. The title, in the bottom right corner, reads: 'A plat of the Mañor of Notton wherein all the Demane groundes are colored with yelowe, The tennantes Inclosures are Colored with grene, the feilde groundes are colored with red, and the freholdes with the Comons and wastes are left white made by Christofer Saxton Anº: Dñi: 1599'. To the bottom left is the line scale measuring six inches, divided into ninety-six perches and numbered at sixteen perch intervals, with the written statement above: 'This Scale contayneth 96 perches', giving a scale of sixteen perches to one inch. Some colouring is retained, but shades are indistinguishable apart from the watercourses which are still blue.

The whole of Notton township is described, with roads, watercourses, woodland, buildings, field boundaries and paling surrounding Notton Park all shown; there is no indication of high ground. Both enclosed and open fields are named together with tenants and freeholders. Land use, i.e. meadow or pasture, is noted on the fields of Michael Wentworth's tenants. Buildings are scattered throughout the township but with a nucleus around Notton Green. Two mills are shown, one called 'Aboldehay mylne' and the other on the Woolley boundary on the stream from Seckar, many years later to be 'modernised' by James Brindley. No land measurements are given.

The written survey of Cold Hiendley, at the Brotherton Library, Leeds University,[51] is on two sheets of paper and headed: 'A Suruey of the Mannor of Coldhieneley made by Christofer Saxton A°: Dm: 160(5?)', with the paper torn away obliterating the last figure. It was probably made in 1605 as G. E. Wentworth states in 1893: 'Mr Francis Woodrove also sold a few farms in Cold Hiendly to Michael Wentworth, and there is also a survey of Cold Hiendly made by Saxton in 1605.'[52] There is a diagonally curved line following the hole which is probably just an extra flourish of the pen as is sometimes found on other surveys, for instance that of Ledston by Robert Saxton. This survey is actually written by Robert Saxton, confirming that Robert acted as his father's assistant at this period. Land measurements are recorded in acres, roods, dayworks and perches, and again the numerical additions are inaccurate. The survey details the holdings of four tenants, the site of the manor, the field land and the common, which together total 341a 3r 3dw 2p (or slightly more when added up by the writer!). Arable, meadow and pasture lands are remarked upon. The survey is clearly written and completely legible.

In November 1599 Michael Wentworth purchased the Woolley and Notton estates from Francis Woodrove. G. E. Wentworth wrote: 'Christopher Saxton, the great mapmaker, surveyed the three manors of Woolley, Notton, and Coldhiendley for Mr Michael Wentworth at this time; and the surveys still remain at Woolley.' He briefly describes the Woolley survey and continues: 'In Notton there were 1477 acres.' This information cannot be deduced from the existing map, so either Wentworth introduced the information from a completely different source, such as the deeds or a later survey, which seems unlikely when he was discussing Saxton's surveys, or, what is more likely, he was then in possession of a written survey of Notton in addition to the map. Unfortunately no survey has survived with the papers later deposited by the Wentworth family at the Brotherton Library or the Yorkshire Archaeological Society. All three surviving documents must have been made at the time of purchase of the estates purely to facilitate Wentworth with a knowledge of his new lands and tenants' holdings.

Ribston; Hunsingore; Walton Head, Yorkshire

At Ribston Hall, north of Wetherby in the West Riding of Yorkshire, are two large maps which are copies of those made by Christopher Saxton in 1599 and 1600. They were copied in 1651 and bear the initials JG, probably for John Goodricke who owned the Ribston estates at that date. There is no trace of the originals, and no supportive documentary evidence.

The first map measures 102cm by 310cm and is drawn on parchment. The title, in Saxton's usual box cartouche, is at the top left and reads: 'A plat of y*e* Demesnes of y*e* manor of Ribston with Walsford & Walkers Ferme Wherin Ribston is coloured Yellow, Walshford with green, Walkers Ferme with Red, and the confines left white made by Christofer Saxton A.D 1599. And coppied out 1651 Riens sans J. Dieu.' It is endorsed: 'SURVEY of y*e* Mannour of Greate Ribston Walshford & Walkers Ferme 1651' in a widely spaced double line box cartouche. The map has a wide double line border, thickly painted in red, with spaces left colourless for the cardinal points in Roman capitals in

English. There is no scale given. The whole map is heavily painted, most of the colour remaining true.

The river Nid is shown across the south of the area mapped; part of Goldsbrough is shown to the west, 'Hopperton Carre' to the north and 'Walshford Moore' to the east. Park paling is depicted surrounding a comparatively small area in the centre of the map. The only buildings shown are situated between the pale and the river, with fishponds nearby. Fields are named and measurements given in acres, roods, dayworks and perches. No trees or high ground are indicated.

One of the most interesting points to be found on this map is an explanatory list of land measurements. Whether this was copied from the original, or added by the copyist cannot be said. It reads:

   16 foote & ½ a Perche
   4 Perches a Daworke
   10 Daworks a Roode
   4 Roodes an Acre
   4 Perches in bredth & 40 in length makes an Acre
   22 Yards in bredth & 220 in length make an Acre
   There is 160 Perches in an Acre

One can argue that as Saxton always used dayworks it would seem likely that he made the note; on the other hand it may have been felt necessary to add the explanation in 1651 as dayworks were so rarely used locally as a land measurement.

The second map at Ribston was originally made in 1600 and is again on parchment. The title, in the top left corner is in a simple single line box and reads: 'A Platt of ye Lordshippe of Hunsingore, greate Cattall & Lune-house, made by Christofer Saxton Anno Dñi 1600 Coppied 165½ Rien sans Dieu ⑂ .' Inscribed on the reverse is 'Survey of ye Mannours of Hunsingore, Cattal Magna &c. – 1651'. The map measures 119cm by 149cm, and has a similar border to the Ribston map. A rather crude line scale, not in Saxton's usual style, is drawn in the bottom right corner; it is simply four lines dissected at intervals and marked at 16, 32, 48 and 64 perches, with the statement '64 Perches' above. It measures four inches, giving a scale of sixteen perches to one inch. There are no dividers. Above the scale is a note of the land measurements, exactly the same as on the Ribston map except that a portion is torn and missing. The map is brightly coloured with all-over washes, but rather crudely executed.

The River Nid again forms the southern boundary of the area mapped, with just Colthorp Church to the south; part of Gelsthorpe Moor bounds the northern area, parts of Lamberton and Whixley the north-east, Little Cattal the east, Cattal Moor the south-east, part of Ingmanthorpe the south-west and part of Walshford the west. Fields are named and in some cases measurements are given in acres, roods, dayworks and perches. Habitation is shown at Colthorpe in the south-west, with a mill nearby on a goit running from the river. Immediately north of the river is the New Hall at Hunsingore which Cromwell's men later demolished; only terracing now indicates the site. The position of Great Cattal is indicated, but no buildings shown. The area shown on this map lies to the east of that on the Ribston map. It is not known why Saxton made visits in consecutive years, 1599 and 1600, unless it was over the new year period of March April when he could have completed the task in one session. Both maps were presumably made as estate surveys.

In 1599 the estate was owned by Richard Goodricke, who, in 1559 had been heavily involved in the Rising of the North. He died in September 1601 and the estate passed to his son, Henry, who was knighted by James I.[53] Henry Goodricke married Jane, daughter of Sir John Savile of Methley, Baron of the Exchequer, who was related to the Saviles of Thornhill and it is quite feasible that word of Saxton's expertise reached Ribston from that quarter.

Sir John Goodricke, in whose time the maps were copied, had taken arms against parliament in the Civil War and as a result his estates were sequestered and he imprisoned in The Tower. In 1646 he paid a fine of £1,200, was pardoned and his estates were restored to him. He retired to Ribston and it was at that period of taking stock that the maps were copied. The maps appear to be true copies of Saxton's originals, as evidenced by the inclusion of the New Hall at Hunsingore which was demolished by Cromwell's army prior to the maps being copied.

Ribston remained in the hands of the Goodrickes until 1833 when the estate was lost in a game of chance. In 1835 it was purchased by Joseph Dent, whose descendants still live there. Ralph Thoresby, the Leeds antiquarian, records in his diary that he visited Ribston on June 8th, 1710. He was 'most courteously received by Sir Henry Goodricke' who showed him various items of interest and 'the autograph and some original surveys of Christopher Saxton's;'. We must presume that Thoresby saw the maps here described.

There are copies of both maps at the Yorkshire Archaeological Society.[54]

Richard Goodricke employed Saxton again in 1600, this time to survey his estate at Walton Head in the parish of Kirkby Overblow, midway between Harewood and Harrogate in the West Riding of Yorkshire.

A written survey survives in Leeds City archives, and is written on a single sheet of parchment endorsed 'Walton Head 1600'.[55] The survey is headed: 'A Suruaye of the Lordshipp of Walton Head taken by Christofer Saxton. 1600.' and is written in an unknown flamboyant hand. The whole sheet is 'framed' by a narrow double line border and although different styles of writing are employed, none is Saxton's. Land measurements are given in acres, roods, dayworks and perches.

The first section lists the 'Demanes' with a pointing hand. Tenants included Anthonye Tayler who occupied Walton Head House with the orchard and a total of 46a 3r 1dw 2p of land. Six other tenants are named and their holdings listed, totaling 241a 1r 1dw 1p. Then follows a list of the farms, again listing tenants, their holdings and acreage. Problems with addition are apparent, the total of the demesnes being written in, crossed through and a second total inserted. In fact the first was nearest correct, being only one perch out. There were seven farms with a total of 111a 2r 2dw 1p; one had a windmill.

Walton Head had belonged to the Johnson family for some time, but like many other Catholic families their fortunes diminished and their lands were forfeited following the rebellion in 1569. Subsequently Henry Johnson was pardoned and his lands restored. One of his daughters, Frances, married Sir Francis Baildon, whose second wife was Margaret Goodricke, sister of Richard Goodricke of Ribston. In 1582–3 Sir Francis Baildon sold Walton Head to Richard Goodricke, and it was he who employed Saxton to make a survey of the estate in 1600.

Richard Goodricke's second son, Lieutenant-Colonel William Goodricke who died in 1662, is described in his will as of Walton Head.[56] The manor eventually came into the ownership of the Lascelles family, Earls of Harewood, in whose archives the survey was recently discovered.

Dewsbury, Yorkshire

Dewsbury Public Library are the owners of a map of that town made by Christopher Saxton in 1600. It is the only town map to have survived, although, of course, towns and villages are shown on some of Saxton's other maps. Dewsbury lies only five miles south-west of Dunningley and would doubtless be well-known to Saxton.

The map measures 56½cm by 41cm. The title, just below centre to the right, reads: 'A plat of the Towne of Dewesbury with the course of the Riuer and waies from Maister Birkbye mill, to the ouer, and nether myllnes of Dewesburie. Made by Christofer Saxton. Anno Domini 1600.', with a neat line embellishment beneath. Below the title, in the bottom right corner, is the scale rule measuring two and a half inches in length, divided into perches and numbered at 16, 32 and 40. Above is written: 'This Scale is 16 pches to ye Inches', with 'ye' inserted above as an afterthought.

The map shows the 'Calder flude' (river) snaking across the sheet from the south-west corner to the north-east, from the mill dam and goit to Brook Hole, where Batley beck enters the Calder. Dewsbury town is shown in the north-east with a profusion of buildings, including the church where Saxton's ancestors are buried. Master Birkby's mill, the manorial mill, was built on Batley beck in the town itself, just north of 'Wakefeilde Way' and east of 'The way from Dewsburi to Stanclif'. The other mills mentioned in the title are at either end of the goit shortcutting a loop in the Calder, just downstream from the mill dam. To the south of the river lies Thornhill which Saxton had already surveyed. In the river to the south of the church is an island, named again by Saxton as 'The Steaner', as was his custom. Various named routes and ways are shown, either as double continuous lines, or double dotted lines. Milnefeilde Laine, running alongside the river, was submerged in part, and Saxton noted: 'Under this in the water was the highway.'

The map was purchased by Dewsbury Corporation in 1899 for £10. It formerly belonged to Sir Thomas Phillips, manuscript No.34556. Unfortunately any papers that may originally have accompanied the map are now divorced from it, and without them it is difficult to discover the exact reason for the drawing of the map. It would seem likely that it was prepared in connection with some legal dispute concerning water rights and the mills. Enquiries and searches have been made locally and at the Public Record Office, but no documentation can be traced concerning a dispute, or indeed giving any clues as to why the map was made. Although Dewsbury is an ancient parish with a long and interesting history, it has been sadly neglected by historical writers. It was in the Manor of Wakefield and part of the Duchy of Lancaster at the time in question, with the Queen as Lady of the Manor.

Hadlesey; Keighley, Yorkshire

Christopher Saxton was engaged by William Cavendish to survey two of his Yorkshire estates, firstly that in Hadlesey in 1600, followed by that in Keighley in 1604. The surveys are both written, in an unknown hand, in a vellum bound book belonging to the Duke of Devonshire. Written on the cover of the book is: 'Survey of the Estates of William Lord Cavendish 1609' and there is an accompanying book of maps which relate to most of the surveys, but unfortunately not to the two by Saxton. The first book contains many surveys, dated up to the late 1620s, almost all of which are by William Senior, 'Senior Professor of Arithmetique, Geometrie, Astronomy and Dialinge' as he describes himself in 1616, or 'Welwiller to the Mathematiques' in 1627.

The two Saxton surveys would appear to have been copied into the volume, probably in 1609, but there is no record of the original surveys delivered by Saxton a few years previously. Both surveys are on paper, the first being entitled: 'A Survey of the Mannor of Hadlesey made by Christofer Saxton A⁰: Dñi 1600.' It commences on page twelve of the volume and is two and a half pages long. Land measurements are described in acres, roods,

Manuscript map of Snapethorpe by Christopher and Robert Saxton, 1601.

dayworks and perches and the total land surveyed was 1,428a 1r 9dw 3p. The pages are clean and legible, but in a difficult script.

Hadlesey is situated five miles west of Selby in the West Riding of Yorkshire, about eighteen miles from Dunningley. The estate had previously belonged to Edward Savile, and was inherited by Sir George Savile, who, in 1591, sold the manor to John Elwys. It changed hands twice before being purchased by William Cavendish.[58] It was presumably on his purchase of the manor that Cavendish employed Saxton to make the survey.

The second survey by Saxton for William Cavendish is headed: 'A Survey of William Cavendishe Esqrs Lands in Kigheley made by Christofer Saxton Anno Domini 1604.' It commences on page fourteen of the volume and is six and a half pages long, with a few lines on the corner of the following page. The total quantity of land surveyed was 2298a 3r 1dw 0p. The script is the same as in the Hadlesey survey, which it follows. Keighley is situated nine miles north-west of Bradford in the West Riding of Yorkshire.

The Keighley estate had come to William Cavendish from his father, Sir William Cavendish, who is recorded as Lord of the Manor in 1540.[59] William Cavendish, whose mother was the renowned Bess of Hardwick, married Ann Keighley, daughter and heir of Henry Keighley of Keighley in 1582. She brought considerable estates in Yorkshire to the marriage, but it is clear that the Cavendish family already owned lands in Keighley prior to the marriage.

Snapethorpe, Yorkshire

In 1601 Christopher Saxton was employed by Thomas Pilkington to map his estate at Snapethorpe, about a mile and a half west of the centre of Wakefield, now part of that city, and almost five miles from Dunningley.

The map, which survives in private ownership, measures 71cm by 76cm. It has the title: 'A PLAT of Mr Pilkinton his landes, belonging to the old Haule, & new Haule, in the parish of Waikfeild, Made by Christofer Saxton. Ano 1601,' contained in an elaborate cartouche, quite uncharacteristic of Christopher Saxton, in the bottom left corner of the map. In style this resembles the engraved cartouches of the printed maps of the period with a decorated border and wooden scrolls. Emerging from the top of the cartouche are two archers with bows drawn aiming at a bird, possibly an owl, perched on a branch. Of greater significance are the initials RS drawn in circles on the bottom of the cartouche, the initials of Robert Saxton, Christopher's son then aged about sixteen; the earliest evidence of Robert assisting his father and who was probably given the task of drawing up this map, although the writing on it is in Christopher's hand. To each side and below the cartouche are further embellishments; to the right a highly coloured compass rose, and to the left the Pilkington family crest of a left-handed mower with a scythe. Above the cartouche is the line scale measuring four inches, divided both horizontally and vertically, being marked in perches and numbered at 16, 32, 48 and 60. It is surmounted by open compass dividers between the legs of which is a scroll, similar to those found on Saxton's engraved maps, containing the words: 'This Scale conteyneth 64 perches.' The scale thus being sixteen perches to one inch. The cardinal points, within the border in Roman capitals in Latin, are in individual ornamental boxes (Plate 16).

The area mapped is bounded on the north by 'The waie from Halifax to Wakefeld', and from north-west to south by an unnamed watercourse (Lights Beck in the north and Lupset Beck in the south). A road traverses the southern third, on the route of the present Horbury Road, and Lupset Hall is shown to the south-east, with part of Lady Savile's land in the south, and William Savile's land in the east. The only other buildings drawn are two small structures named 'The Scite of the old haule with the orchard & lathe', and a tower just south of the Flanshaw Lane junction with the road to Halifax. The rest of the map shows named fields, with most of the tenants given. There are no trees or indication of high ground. The map originally had the boundaries painted yellow, now faded, but is otherwise uncoloured.

In 1601 the manor belonged to Thomas Pilkington, whose ancestor Sir John had been granted the manors of Snapethorpe and Lupset in 1474.[60] In the mid sixteenth century Snapethorpe had been let, but Thomas Pilkington was in occupation at the time of Saxton's employment, and Lupset then belonged to a junior branch of the Saviles of Thornhill. In 1603 Pilkington purchased Stanley Hall to the north-west of Wakefield, where he then lived until his death in 1611. The original Hall at Snapethorpe, built in 1477, was called Pylkington Hall; by the sixteenth century it was known as the Old Hall, and in recent times was known as Snapethorpe Hall.

The crest illustrated is that of 'a mower with a scythe, habited per pale argent and sable' and was in use as early as 1424 on the seal of Sir John de Pilkington. The family motto 'Now Thus, Now Thus' is believed to have originated after the battle of Bannockburn in 1314 in which Pilkington participated. The legend told is that Pilkington, in fleeing for his life after the battle, took up a scythe to disguise himself as a rustic mowing. Unfortunately he held the scythe backwards way and was thus discovered. The mower in the crest is half in black, half in white, i.e. half rustic and half in armour, depicting the alternation of flight, or concealment, and recovery therefrom.[61]

The northern part of the estate is now known as Lupset Estate, it was purchased by Wakefield Corporation in 1926 and a housing estate built there. Snapethorpe Hall was demolished in the mid 1970s.

The map of 1601 was made as a general survey of the estate and the tenants. As measurements are not noted there may originally have been an accompanying written survey.

Burley, Yorkshire

In 1602 Christopher Saxton surveyed the Lordship of Burley, which lies between Otley and Ilkley in the West Riding of Yorkshire, fourteen miles north-west of Dunningley as the crow flies.

A written survey survives at the Yorkshire Archaeological Society in the Middleton Collection.[62] It is written by Saxton and is on two sheets, being headed: 'A Suruey of the Lordship of Burley made by Christ: Saxton Anᵒ Dñi: 1602.' Measurements are given in acres, roods, dayworks and perches. The survey commences by listing Mrs Calverley's lands and 'the Scite of the Maner of Burley wᵗʰ the orchard Mylne and Mylnehill, and veuer close', totaling 447a 2r 8dw 3p. Seventeen tenants are then listed, with the land in their occupation. Once again the numerical additions are inaccurate.

At the beginning of the seventeenth century the Lordship of Burley was held by the Middleton family. Apparently Sir Peter Middleton was a benevolent landlord and upon deciding to sell the manor gave the poor tenants their houses, yards and free common for ever, and sold the farms to the tenants at easy rates. The manor was sold by Sir Peter Middleton of Stockeld to Francis Pulleyn of Sicklinghall in 1622.

Without evidence to the contrary, it is presumed that the survey of 1602 was commissioned as a general view of the lordship. The survey is transcribed in full by Fordham.[63]

Ashwell, Rutlandshire

A copy of a map by Christopher Saxton of the Manor of Ashwell, in Rutlandshire, made in the early eighteenth century survives in the Dawnay archives at the North Yorkshire Record Office.[64] The original was made in 1604.

The copy measures 46cm by 59cm, and has a double-lined title box adjoining the bottom border to left of centre, which is painted red. The title reads: 'A Platt of ye Mannor of Ashwell wherein all ye Demesnes Inclosed are coloured with Yellowe. The Lords Lands not Inclosed wth. Red The Freeholds wth. Greene The Land appointed for y^(e) Tenants to be Inclosed wth. another Red and ye Confines left White originally made By Christopher Saxton Anno Dom: 1604.' To the left is written: 'This was a copy made by Joseph Pepper of Stamford'. The scale rule, at the bottom right, gives a scale of forty perches to an inch and is surmounted by a pair of open dividers, rather more delicately drawn than Saxton's usual portrayal. In the top right corner is a highly coloured sixteen point compass rose. There is a narrow double line border, coloured red in portions only. There are no cardinal points.

The map, which is now rather worn, shows the Manor of Ashwell, about three miles north of Oakham. A collection of buildings is shown in the centre of the map but much of the detail, including the writing in that section, is largely obliterated. The remaining area mapped is field land, with some field strips. Occupants, and some field names, are given. Measurements are in acres, roods and perches only, and may be a later addition. Trees are drawn in 'The Leys above Powde hill' only. Roads are depicted by either single dotted lines or double dotted lines. There is no indication of any high ground.

A Joseph Pepper, schoolmaster of Stamford (1703?–32), was active as an estate surveyor in Northamptonshire. There was also a man named Pepper who was agent to Lord Exeter at Burghley in the late eighteenth century, but it seems that the former is the more likely candidate as the copyist of Saxton's map.[65]

At the time Saxton drew his map, the manor of Ashwell was held of the Earl of Rutland's manor of Belvoir by Francis Palmes. The Palmes family sold the estate in 1699 to Bartholomew Burton, whose granddaughter Lora married John Dawnay, Viscount Downe, in 1763. It remained in that family until sold by the 8th Viscount Downe in the second half of the nineteenth century.[66] The copy of Saxton's map must have remained with the estate papers acquired by the Dawnays when they purchased the manor. All attempts to trace the original map have failed; possibly it was discarded when the 'new' copy was made by Pepper. It was probably made as a survey of the estate at a time of enclosure.

Spofforth, Linton, Yorkshire

Two estate maps by Christopher Saxton survive in the archives of the Duke of Northumberland at Alnwick Castle.[67] Both maps are of the Manor of Spofforth, which is between Wetherby and Harrogate in the West Riding of Yorkshire, some fifteen miles north-east of Dunningley. One map was made in 1606, the other in 1608. There is also evidence at Alnwick Castle that Saxton surveyed Linton, to the south-west of Wetherby, on one of his visits, but no map or survey of Linton survives.

The first map is the smaller of the two, measuring 62cm by 109cm. The title, just above centre on the right, reads: 'A plat of the Mannor of Spoford wherein all the demanes and wood groundes are Colored with yelowe, the Inclosures and Feildes Belonging to the towne are Colored with grene, the Comon with red, and the confines Left White. made by Christofer Saxton An°: Dñi 1606.' The scale rule, in the bottom right corner, measures two inches. It is marked in perches and numbered at 20, 40, 60 and 80, giving a scale of forty perches to one inch. Below is the statement 'This Scale Conteyneth 80 perches' and the whole rule is surmounted by a simple pair of open dividers stretching to each end of the rule. In the top left corner is a rectangular double line box cartouche which is empty, suggesting that the map is unfinished; presumably the box was intended to contain explanatory notes.

The map itself is in very good condition. It shows the Manor of Spofforth, which lies to the south-west of Ribston which Saxton surveyed in 1599. Spofforth 'towne' is shown to the north-east of the map, with its manor house, church and comparatively large number of houses clustered together. Two mills are drawn to the north of the town. Much of the manor was wooded, but the large central area, almost half the manor, was empaled parkland, which contained very few trees. The land is all named, but no tenants are given. Roads are shown, both by parallel continuous lines and double dotted lines. Field gates are drawn as the letter N across the line of the boundary, road, or whatever. The field names are sometimes followed by the letter A, P, M or W, signifying whether the land was arable, pasture, meadow or woodland, although the description is sometimes written in full. In addition there are a variety of other symbols: + ++⊙△ ◇ * but the significance of these is not known, unless it was to indicate the quality of the land. In a few instances acreage is given, but does not include dayworks. The map is colourwashed all over; the borders, scale rule and dividers are painted a pale yellow-brown, now faded; the roofs are red and the watercourses and 'the Stanke' are blue. The latter is a lake to the north of the manor house with 'the Howse milne' built on its northern perimeter. There is no indication of high ground.

The second map of Spofforth is larger, measuring 119cm by 149cm. The title, in the top left corner, reads: 'A Plat of the Mannor of Spoforde Wherein all the Demanes and Wood groundes are Colored With Grene, the Inclosures and Commonfeilds with red, the Common with yelowe and the Freholders Left White. made by Christofer Saxton A.nno Dñi: 1608.' The scale rule, bottom left, measures three inches, and is divided into perches numbered at 20, 40 and 60, with the statement 'This Scale Conteyneth 20 perches to an Inche', with simple open dividers surmounting the whole rule.

The area mapped is the same as in 1606, but far greater detail is given on this map. Tenants' names and acreages, in acres, roods and perches only, are given on even the smallest holding, which in Spofforth town and vicinity tends to make it a very crowded map and consequently illegible in parts. Freehold and copyhold land is noted, and the same letters A, M and P are used to indicate arable, meadow or pasture land; W for woodland is not used. The symbols ⊢ + ++ are found on some fields, again without explanation; Batho suggest that they may signify relative quality.[68] Paling, roads and bridges are shown as on the 1606 map. This map too was originally colourwashed all over, but this has now faded.

In both Spofforth maps adjacent areas are named and it is perhaps relevant to note that to the south-west of the manor is written 'Part of Stockeld Park Mr Middletons', and adjoining, further south, is 'Part of Siclinhall Mr Vavisors'. Mr Middleton it will be remembered, employed Saxton to survey Burley Manor in 1602, and the Vavasors had relatives in Baildon for whom Robert Saxton worked in 1610.

Spofforth Manor was held by Henry Percy, 9th Earl of Northumberland, in 1606 and 1608, but who was at the time imprisoned in The Tower following the involvement of his cousin Thomas Percy in the Gunpowder Plot. As an absentee landlord his estates had become mismanaged and at this period the Earl was seriously trying to take stock and reorganise administration. In 1609 the Earl wrote his 'Advice to His Son', in which he said the first principle of estate management was to 'understand your estate generally better than any one of your officers'. He claimed 'I have so explained and laboured by books of surveys, plots of manors, and records, that the fault will be your own, if you understand them not in a very short time better than any servant you have. They are not difficult now they are done, they are easy and yet cost me much time and much expense to reduce them into order. . . .'[69]

In addition to the maps it is most interesting to find an account roll at Alnwick Castle which contains a reference to Saxton's survey of Spofforth. He was apparently paid £16 4s 9d for surveying the Manor of Spofforth, but it is not indicated whether this was for one visit or two.[70] What is perhaps a more exciting discovery is the fact that Saxton was also paid for surveying Linton, another of the Percy estates, for which he received £11 4s 5d. On the same account roll is an entry 'for Charges in fetchinge Mr Saxton to S^{r}vey the Manno^r of Spofforth & Lynton xij^s vj^d' (12s 6d).

The account roll includes payments to other surveyors employed, notably to John Thewe for surveying the 'Courseing Parke at Leconfield', and land at 'Scorbrough Cheryburton Hasell and Beverley'; to John Judson for 'money by him paid for the S^{r}veying of his Lo: Lands in Wilbfosse and Catton'; and to William Lampe 'for S^{r}veying of Scorbrough Common & the Intacke there, and setting forth his Lo: a fixt pte to be inclosed, and for S^{r}veying of Catton Old hagge'. None of these surveyors incurred charges for being fetched, and all were paid less than £2 for their work; of course it cannot now be ascertained what was involved in each case, but Saxton certainly received considerably more than any of the others for his work.

Unfortunately this is the only reference to a survey or map of Linton to have come to light. Neither are at Alnwick Castle, and it seems most probably that whichever it was does not survive.

River Swinden, Lancashire

In 1606 Christopher Saxton made a return visit to Lancashire, this time on Duchy of Lancaster business, in connection with disputed water rights. Saxton was employed to map the course of the River Swinden and its tributaries over a distance of a mile and three-quarters; the area lying due east of Burnley near the Yorkshire border.

In addition to mapping the area, Saxton gave evidence to the appointed commissioners including details of his age.

The map by Saxton survives at the Public Record Office, is on three sheets of paper and measures 41cm by 81cm.[71] The title, in the top left corner, reads: 'A perfect plat of the millnes and Water Courses in question betwixte John Towneley Esquire. plt: and John Parker Gentleman and otheres Defendantes, made by Christofer Saxton the . 23 . of Aprill. An^o: Dñi: 1606.' The scale rule, in the bottom right corner, measures two and a half inches and is marked in perches, with '40 perches' written at the end. Beneath is the statement 'This Scale Conteyneth one furlong', giving a scale of sixteen perches to one inch. The rule is surmounted by open dividers.

There are two explanatory notes on the map. Firstly, 'The distance betwixte the milnes is thre(e)quarteres of a mile in a righ line, and by the compas of the water a mile.'; and secondly, 'Note that the distance betwixte M^r: Parker milne and the heade of the Riuer Swinden is two mile and a quarter, and hath diueres springes falling into him as Renalde well howlden Clugh little Coole Clugh Greate Coole Clugh and otheres'. 'Swinden brooke' is shown flowing from east to west across the map, still highly coloured, with several tributaries and goits. Extwistle lies to the north of Swinden brook and Worsthorne to the south; land belonging to Mr Towneley and Mr Parker is noted. There are no proper roads drawn, but a single line, possibly a track, is shown leading to various houses. Mr Towneley and Mr Parker had adjacent houses at Extwistle and 'Mr Parker milne' is shown nearby at the foot of a dam fed by a goit from the brook where Saxton marks 'The Calle or were'. A mile downstream, on a dam at the foot of a long goit, is the 'Ould milne' on Mr Townley's ground. There are two breaches shown in this goit, and at the head is 'The marle banke worne away and trees worne downe'. Some trees are drawn along the riverside, but there is no indication of high ground although the area is very hilly.

John Parker was a substantial freehold tenant of John Towneley, the complainant in the case. While the Towneley family were away, Parker had erected a mill fed from Swinden Brook about a mile upstream from the Old Mill belonging to Towneley. The resultant dispute came before the Duchy of Lancaster Court and the commissioners appointed took evidence at Burnley on April 23rd, 1606, the same date as on Saxton's map. According to one of the deponents, the dam of the new mill was about three times as large as that of the old mill and took eight hours to fill. Furthermore, the new mill turned the watercourse and as a result it was not possible to use the old mill to grind corn, as formerly, while the new one was operative.

Saxton had been engaged to draw a map of the watercourse to determine to what extent Swinden Brook was fed from springs and wells below the new mill to substantiate or refute this. He was called before the commissioners and submitted the following deposition: 'Christopher Saxton of Dunningley in the countie of Yorcke gent. aged lxiiij yeres or thereabouts sworne and examyned to the xviij[th] Interr deposeth and sayeth that he this depon[t] haith seene and veiwed the water or Brooke of Swinden Runinge betwene the Comp[ts] auncyent mylne and the newe mylne called Mr Parkers mylne and haith measured the same. And sayeth that he this deponent did not see or pceyve any well springes or Runnells of water w[ch] doe Come or Rune into the saide water or Brooke of Swindon betwene the saide mylnes but onlie suche as he This deponent haith expressed and sett downe in one Platt by him this deponent made and nowe shewed and delyvered unto the saide Comission[rs] at the tyme of this his examynacon, whereunto he this deponent referreth himselfe, upon w[ch] platt the said comyssions or some of them have Indorced theire names.'[72]

This is the second occasion on which Saxton declared his age, the other being in evidence connected with the Isle of Axholme dispute. Unfortunately the two ages given do not correspond. In September 1596 he said he was fifty-two 'or thereabouts', which would make him sixty-two years old in September 1606, whereas he states in April 1606 that he was aged sixty-four yeres 'or thereabouts'.

Liversedge, Yorkshire

A map, or survey, of Liversedge, east of Halifax in the West Riding of Yorkshire, reputedly made in 1608, would appear at present to be the last commission carried out by Christopher Saxton. Unfortunately the result is lost.

Frank Peel in his history of the Spen Valley wrote extensively on the history of Liversedge, and quoted extensively from papers and notes assembled in the sixteenth century by an antiquary named John Hanson.[73] Peel states that these papers were discovered in the Bodleian Library at Oxford, where they had remained undisturbed for more than three centuries until he discovered them. Unfortunately Peel gives no reference by which the collection can be identified today and the papers cannot now be traced. However, the extracts and descriptions given by Peel would appear to be an authentic record, and in the absence of the original the information below is taken direct from Peel's transcripts and interpretations, as being the only record of Saxton's involvement.

In 1573 the Queen granted Liversedge to Edward Carey, Steward of the Manor of Wakefield and who lived sometime at Sandal Castle. He was much favoured at Court, his father being the Queen's cousin. Carey rarely visited Liversedge and used his estate there merely as a source of revenue, causing friction amongst the tenantry. Edward Carey died in 1608 and James I 'sent down a commission to survey the Liversedge Manor before he transferred it to Sir Phillip Carey the next in succession'. The commissioners went armed with a list of 'articles to be inquired of', dated October 17th, 1608, which contained sixteen items. A jury composed of tenants certified their answers at Liversedge two days later. Their answer to the second article concerning the state of the mansion house and demesne lands is lengthy, but explains that the manor of Liversedge was divided into three hamlets, one of which was known as Liversedge Robert (or Robert Towne), wherein was situated Liversedge Hall. Speaking of the hamlet, the tenants state their uncertainty as to rents and, 'neither what quantity of acres the demayne lands do contayne, but they be informed that the said demayne lands be surveyed by Christopher Saxton an experte surveyor', which survey was in the hands of John Jackson, officer and receiver to Sir Phillip Carey. Later in the same answer, this time concerning 'Tomsons Ferme' in Long or Great Liversedge, the tenants are listed but cannot say what acreage each held, stating: 'The contents of the acres of which tenement and lands they cannot certify, but refer themselves to the said survey.'

A search in all likely sources at the Public Record Office has been made in an attempt to trace further references to Saxton's employment, but without success. In 1614 Christopher's son Robert was employed to make a further survey of Liversedge, see page 131.

ADDITIONAL NOTE

In 1950 the existence of a map of Belfast Lough, Ireland, reputedly by Christopher Saxton and dated 1569, was reported in a posthumously published article by Edward Lynam, Superintendent of the Map Room at the British Museum.[74] This map, if genuine, was an exciting discovery in that it represented the only known work by Saxton prior to his great national survey. A copy of the map is in the Public Record Office in London, in Lord Burghley's collection, and this bears no resemblance to Saxton's later known work.[75] The map supposedly found in Ireland, Lynam reported, had no title but was endorsed 'Baie of Cragfargus' and signed 'Christopher Saxton after Michal Fitzwilliams 1569'.

Lynam unfortunately died before actually seeing the map, having taken the information concerning the find from an article in the *Belfast Telegraph* written by a Mr C. J. Robb in 1949.[76] Since that time J. H. Andrews of Trinity College, Dublin, an authority on early Irish maps and cartographers, has made exhaustive enquiries but has been unable to trace the map. Dr Andrews, in his paper 'Christopher Saxton and Belfast Lough', details his lengthy investigations both in Ireland and Australia in an attempt to trace the map in question, and of proving its authenticity or otherwise.[77] It became increasingly clear that a genuine Saxton map had never existed and the most likely explanation put forward is that a modern reproduction of the Public Record Office map had been embellished with the endorsement and attribution.

A second (probable) misattribution, this time concerning a survey of the Outwood at Wakefield was referred to by G. E. Wentworth in his history *The Wentworths of Woolley*.[78] In a footnote concerning Christopher Saxton, Wentworth says that 'In a suit in the Duchy of Lancaster Court affecting the rights of tenants of the Wakefield Outwood, depositions were taken at Doncaster, in September 1638, and among the witnesses examined was Thomas Somester of Wakefield, gent. aged 84, who testified to his having been commissioned with "one Mr Saxton, a surveyor of land", to survey the Outwood in the time of James I.-J.C.C.'

The deposition in question, submitted by Thomas Somester, aged 'eightie fower years or thereabouts' goes into great detail about the state of the Outwood.[79] Somester says he was a Commissioner for the surveying of the Outwood, but nowhere in his deposition is there any mention of Christopher Saxton. Wentworth quotes 'J.C.C.' as his source for this information, probably J. J. Cartwright, M.A. Cantab, of the Public Record Office. Whether either of these historians misread the document, or whether there was another court case in the same year at which Somester testified cannot be said, but no evidence can now be traced connecting Saxton with the survey of the Outwood.

Chapter 9

Robert Saxton's manuscript maps and surveys

TABLE 10

Robert Saxton's manuscript maps and surveys

Date	Place (all in Yorkshire)	Map or Survey	Location	Page
1607	SANDAL	S	Pierpont Morgan Library	122
1608	WAKEFIELD OLD PARK	S	Leeds City Archives	123
1609	FARSLEY	M	Yorkshire Archaeological Society	123
1610	AIRMYN	—	Evidence only	124
1610	BAILDON	M	Yorkshire Archaeological Society	126
1610	ELLINGTON MOOR	—	Evidence only	127
1611	TANSHELF	M & S	Public Record Office	127
1612	ESHOLT	S	Yorkshire Archaeological Society and Leeds City Archives	129
1612	THORNHILL	S	Nottinghamshire Record Office	129
1613	MANNINGHAM	M & S	Public Record Office	130
1614	LIVERSEDGE	M	Evidence only	131
1615	HATFIELD	M	Public Record Office	132
1616	LEDSTON	S	Leeds City Archives	133
1616	SUTTON	S	Leeds City Archives	134
1617	OAKWORTH	M	Public Record Office	134
1618	WENSLEYDALE	—	Evidence only	136
1619	GOMERSAL	M & S	Yorkshire Archaeological Society	136

Sandal, Yorkshire

Robert Saxton's first independent engagement was in 1607 when he surveyed land in Sandal immediately to the south of Wakefield, six miles from Dunningley. The written survey, now in New York, is headed: 'A Survey of diuers groundes in the owldfeillde and in Sandall feilde Ano dni 1607 made by Robarte Saxton:'.[1] It is not in Robert Saxton's hand. Measurements are given in acres, roods, dayworks, perches, and occasionally feet.

Robert Saxton surveyed a total of 55a 2r 9dw op ½ft. He firstly lists holdings in the Owldfeild close, stating whether they are arable, meadow or pasture. Then follows a list of the furlongs in the Oldfeild, 'all wch lands butt on Sandall more on the south And on fallyngs on the north'. 'Lands in Sandallfeilde' follow, butting on the Killnehill Furlong, Pugnall shutt, Rughley hill and Parke shutt, with occupants named and acreage held. There is a note at the end explaining that 'iiij pearches make a Dawarke, And tenn Dawarks make a roode And ffower Roodes an acker'.

The manuscript was formerly owned by Sir Thomas Phillips and sold by Sothebys in 1903, when Maggs Bros. of London purchased it, together with the survey of Ackworth by Christopher Saxton, for £1 11s 0d.

Sandal was in the Manor of Wakefield and in 1607 was part of the Duchy of Lancaster, then held by the crown. Presumably this survey was made as a general survey of land and tenants in the two fields.

Wakefield Old Park, Yorkshire

In 1608 Robert Saxton was employed to survey the Old Park at Wakefield. The written survey, at Leeds City Archives, is on a single sheet of paper, headed: 'A Survey of the ould Parke of Wakefeild made by Robert Saxton a⁰ Dñi 1608.'[2] It is not, however, in Robert Saxton's own writing, which was similar in many ways to his father's. The survey lists the tenants, their holdings and the acreage of land held, given in acres, roods and perches only. There are eleven tenants named in the survey, only one of which, Edward Copley, had a house and garth, no other buildings being mentioned. Mr Savill held the largest portion of almost 70 acres. The total acreage of the park comprised 719a 3r 34p, 341 acres of which were common fields.

The Old Park at Wakefield was bound by the river Calder on the east and south and contained the East Moor, Park Hill and Wind Hill, extending to Lee Moor and the Outwood on the north and west, the paling surrounding it being originally three miles in length. It had been an ancient hunting ground of the Warennes. An interesting description of the Old Park written in 1574 is contained in a report to Lord Burghley by Sir Thomas Gargrave of Nostel, who held the Old Park by Letters Patent dated 1565. Gargrave states that he had surveyed and measured the park which was 'a busshye and barran ground', with no timber suitable for repairing the pale or lodges, but only rotten and decayed trees. The pale was so decayed that it would not keep in the deer and the river had washed away twelve acres of the best ground.[3]

In default of a male heir, the Old Park at Wakefield was granted to Sir Richard Gargrave, old Sir Thomas' grandson, to be held as of the Duchy of Lancaster, by grant dated May 30th, 1608.[4] Only a month later, however, on June 29th, 1608, the Old Park was conveyed to John and Daniel Cowper who in turn conveyed it, in 1613, to Lionel Cranfield who became Lord Treasurer. The Old Park formed part of lands exchanged by Cranfield with Sir Arthur Ingram of Temple Newsam in 1624, and the survey by Robert Saxton was discovered in archives deposited by a Leeds firm of Solicitors who had been employed by the Ingrams of Temple Newsam.

As Robert Saxton's survey is simply dated 1608 it is impossible to state categorically by whom he was employed. It could have been by the Duchy of Lancaster, by Sir Richard Gargrave, or the Cowpers. The fact that the surviving survey handed down to the Ingrams is not in Robert Saxton's hand, suggests that it may be a copy, though contemporary, and that the original may have been sent to the Duchy Court.

Farsley, Yorkshire

The following year, 1609, Robert Saxton was called upon to draw a map of disputed lands at Farsley, three miles north-east of Bradford in the West Riding of Yorkshire, and seven miles north-west of Dunningley.

The map, now at the Yorkshire Archaeological Society, measures 58cm by 84cm.[5] It has the title, in the top left corner, 'A Plat of the Common in Question betwixte the Kinges Warde and Mr: Oldefeild Wherein the Comon in Question is Colored with Grene ʒ Farseley Inclosures with red the Incrochementes with yelow and the Confines Lefte White maide by Robert Saxton A⁰n: Dm̃: 1609.' The scale, to the bottom left, comprises a very simple pair of open dividers surmounting a segmented rule, three inches in length, marked off at 20, 40 and 60, with the statement 'This Scale is 20 perches to the Inch'.

The map shows Preistethorpe and Wadlands to the north, Farsley, as far as Rickershay beck which formed the township boundary to the east, Bulwell Sike, the watercourse forming the southern boundary of the township to the south, and Bullewell to the west – approximately half the township of Calverley. Most of the buildings are situated in

Preistethorpe, Farsley and at Wadlands, which is erroneously shown to the south-east of Priestethorpe Green. The commons and enclosures are named, with occasional comments such as 'The Intakes Inclosed by Mr Caluerley'. The dominant features, however, are the roads and watercourses and their respective diversions, made by Mr Owldefeilde (Oldfield). Several coalpits are shown, drawn as small circles with a dot in the centre. There is a note: 'Here was the limepittes' in Farsley. No trees or high ground are indicated.

This map must have been one of the lesser results following what became known as the Yorkshire Tragedy. The Calverley family had lived at Calverley Hall for many generations, with Walter Calverley, his wife and three infant sons in residence in the early seventeenth century. It was in 1605 that Walter Calverley, a thorough wastrel, committed the sin of murdering his two oldest sons, stabbed his wife (from which she recovered), and rode off to slay his youngest son, Henry, then but a few months old, and with his nurse. Calverley was apprehended before reaching Henry and was taken before Sir John Savile at Howley Hall. At his trial Calverley refused to plead in an attempt to preserve his estates intact for his heir, and died, as was the custom for those refusing to plead, by having heavy stones placed upon his chest until he expired.[6]

Walter Calverley's widow, Phillippa, married Sir Thomas Burton and Henry Calverley survived to inherit his father's estates, but, being a minor, was made a ward of the crown. Sir Thomas Burton, as Henry's guardian, had control of Henry's affairs, but abused his trust and enriched himself at Henry's expense. However, it is apparent from the existence of Robert Saxton's map and accompanying papers that legal problems arose early in Henry's life, for a case came before the Court of Wards concerning disputed land on the common of Calverley and Farsley. The case was between Henry Calverley and Mr Oldfield and concerned encroachments on Farsley Common, and was followed by a similar case against Mr Barton. The Calverleys appear to have bought out the opposition after lengthy evidence had been obtained from local inhabitants.[7]

Airmyn, Yorkshire

It appears that Robert Saxton was employed to survey the manor of Airmyn in the parish of Snaith in the West Riding of Yorkshire in 1610. The survey itself cannot now be located, and the only reference to it is contained in an article written by the Halifax antiquarian E. J. Walker.[8] In the article Walker describes the establishment and early history of the Free Grammar School of Queen Elizabeth at Heath, in the township of Skircoat, now part of Halifax. Walker obviously had access to a collection of papers now lost, and gives an interesting and seemingly authentic account of legal battles to claim a bequest in favour of the school, during which Robert Saxton was employed.

However, in consulting a modern writer, Anthony F. Upton, in his biography of one of the principal characters concerned, Sir Arthur Ingram, certain discrepancies appear, which, without access to Walker's source material, are impossible to clarify.[9]

The school at Heath was established by Letters Patent in 1585 and there were various benefactors, including Brian Crowther of Halifax who bequeathed an annuity or yearly rent of £20 to the Governors of the school from his portion of the manor of Airmyn. In 1606, in accordance with Crowther's Will, the Trustees tried to claim the annuity, but such were the fraudulent practices of one of the other owners of the manor, Edward Waterhouse, Lord of the Manor of Halifax, that the annuity was not paid. There were several cases of litigation, and according to Walker, the matter was referred to Arthur Ingram, Esq. of London, who, on September 4th, 1610, decreed that Mr John Midgley and 'Mr Saxton, the younger, the surveyor', with the assistance of William Taylor, Bryan Richardson, Hugh Worrell and Thomas Baylye, should enclose 200 acres of the common,

Manuscript map of Baildon by Robert Saxton, 1610.

'100 for the cottagers and 100 for the farmers in the said manor; that threescore and seven lands in the tofts be equally allotted to the cottagers at the rate of 5s for every two lands; and 24 acres of meadow, at 6s 8d per acre besides the King's rent . . .'.

John Midgley was probably the attorney of Headley, in Bradford-Dale, one of the foremost legal practitioners of his time in the West Riding, frequently engaged in township affairs and as an arbitrator in local disputes.

Upton, however, having studied the Ingram papers at Leeds City Archives, gives a slightly different version, and does not mention Robert Saxton's employment, which, as neither the survey, nor any specific reference to it, is in the collection, is understandable. Upton, quoting from material in the Ingram papers, says that in July 1608 Waterhouse sold his half of the manor of Airmyn to Ingram, who paid half the purchase price down and agreed to settle the balance in the next three years, providing Waterhouse could prove satisfactory title. There followed more haggling and certain conditions were imposed upon Waterhouse who was asked to clear two annual charges bequeathed by Crowther, one being the £20 annuity to the school. In August 1610 agreements were signed giving Ingram the whole manor and in May 1611 formal release was given. This latter was contested for many years.

Ingram was a very astute London businessman, often acting as negotiator between the court and wealthy city men, or being involved in money-lending, and the legal hassles over Airmy are typical of his dealings. At this period he was building up his estate, particularly in Yorkshire where he was later to settle.

In the articles of agreement, dated August 13th, 1609, between Waterhouse and Ingram in which half the manor was conveyed, it is stated that Waterhouse, at his own expense, had 'to procure and cause to be made a true surveye of all the said mannor and grounds' to ensure equal division and partition.[10] This does not appear to relate to the same document referred to by Walker, but in the absence of the survey itself one cannot be sure.

Vicar Favour of Halifax (who wrote Christopher Saxton's 'epitaph'), was one of the Trustees of the school and he eventually arranged the sale of the Crowther annuity and with the proceeds purchased lands near Halifax.

Baildon, Yorkshire

In 1610 Robert Saxton was employed to map the Commons of Baildon, some four miles due north of Bradford, virtually between Priestthorpe, mapped earlier by his father, and Esholt, mapped two years later by himself. Baildon lies twelve miles from Dunningley as the crow flies.

The map, now at the Yorkshire Archaeological Society, measures 44cm by 58cm.[11] The title, in the top left corner, reads: 'A Plat of Mr Vauisor Comon and M^r: Bayldon Common wherein M^r: Vauisor Common is Colored with grene M^r: Bayldon Comon with yelow, the meane Comon M^r: Ffittwilliams Comon and the Confines lefte White maide by Robert Saxton A^o: Dni: 1610.' The scale, in the top right corner, has a pair of very plain open dividers surmounting a measured segmented rule three inches in length, numbered at 40, 80 and 120, with the word 'perches' following. Above the rule is written: 'This scale Conteyneth 3 furlongs' which gives a scale of 40 perches to an inch. (See Plate 17.)

The map basically describes the commons and inclosures of Baildon, together with their respective owners. Part of Bingley parish is shown to the north and the village of Baildon to the east. There is a 'cole pithowse' and three groups of coal pits on Mr Vavisour's Northwood Common, and a group of four pits on Mr Bayldon's Glovershay Common. Buildings, roads, trees and watercourses are shown, and a roughly sketched hill, Walker foulde crag, is drawn as a sugar-loaf hump. Merestones are drawn as rough triangles along the edge of Northwood, two of which are additions by a later hand. The commons are coloured as described in the title; the border, scale and cartouche are yellow, the watercourses blue, the roads brown and the roofs dark blue. Robert Saxton's map is reproduced by W. P. Baildon in his book *Baildon and the Baildons*.[12]

J. T. Cliffe writes that at the beginning of the seventeenth century there were three gentle families with manorial rights in Baildon: the Fitzwilliams of Bentley, the Baildons of Baildon and the Vavasours of Weston. Several years previously (to 1610) William Baildon and Gervase Fitzwilliam had concluded an agreement whereby the profits of the coalmines (excepting the North Wood Colliery) were to be divided equally between them. 'The Meane Common belonging to Bayldon', shown as the largest area in the centre of the map, belonged jointly to all three manors and could only be inclosed by joint action of all the lords and their tenants. Sir Mauger Vavasour had already begun to sell his Baildon property and Gervase Fitzwilliam followed suit a few years later.[13]

Ellington Moor, Yorkshire

It is recorded that Robert Saxton surveyed Ellington Moor, to the north of Masham, lying between the River Ure and Swinney Beck, in the North Riding of Yorkshire, some thirty-five miles from Dunningley.

No trace of either a map or survey can now be found. The reference was made by Clifford Whone in a paper on *Christopher Danby of Masham and Farnley*, from which the information below is extracted.[14]

Christopher Danby had been the ward of Robert Cecil, eldest son of Lord Burghley, had married Frances Parker, sister of Lord Monteagle discoverer of the Gunpowder Plot, and had inherited large estates in the West and North Ridings of Yorkshire. A scheming steward endeavouring to feather his own nest caused Danby considerable trouble, not least of which was between Danby and his wife. After a brief separation the two were reunited to live at Farnley Hall to the west of Leeds, and in 1608–9 Danby commenced renovations at Leighton Hall, to the west of Masham, one of his other properties. 'A little later' writes Whone, 'Robert Saxton, the son of the famous Christopher Saxton, did considerable surveying on Ellington Moors, but it is not apparent to what purpose.'

Without the evidence of the map or survey, or any further documentation, the reason for Robert Saxton's employment cannot now be ascertained. It could have been to establish what lands Danby owned, or may have been connected with enclosure, or some dispute; Danby's life was dogged by disputes in many spheres.

Christopher Danby's uncle was Michael Wentworth of Woolley who had employed Christopher Saxton in 1599, and Danby was also well acquainted with Sir John Savile of Howley Hall, either of whom may have recommended Robert Saxton to him.

Tanshelf, Yorkshire

Robert Saxton's next commission was to survey and map the manor of Tanshelf, now part of Pontefract, in the West Riding of Yorkshire, and some ten miles from Dunningley. The resultant documents are now at the Public Record Office.[15] Tanshelf was, until this century, a separate manor extending to the west of Pontefract town centre.

The map, measuring 67½cm by 71cm to the outer borders, has the title to the left side a third of the way down the sheet. It reads: 'Demaines of the' (written above) 'A Plat of the Mannor of Tanshelf Colored with Grene the Wāst ground Colored with Red and the Confines Lefte white maide by Robert Saxton. Anno Dm̄: 1611.' The scale rule, surmounted by a very simple pair of open dividers, measures three inches in length, is seg-

A Survey of the Lordshipp of Esholte taken by Robert Saxton Ano: Dm: 1612.

	Acars	Roodes	Perches
Inprimis the scite of the howse with the over orchard	1	1	0
The olde orcharde	0	2	30
The horse coppye	1	1	17
The howme meadowe	16	1	3
The far westfeilde arable	17	2	23
The midle westfeilde arable	12	1	4
The nar westfeilde pasture	27	1	27
The hye roodes pasture	25	0	33
The hye howme meadowe	20	1	11
The hye howme ing meadowe	7	3	12
Stranglefroyth close pasture	25	1	28
The milnefeilde pasture	20	3	26
The long roodes pasture	21	2	30
The Common ing meadowe	13	1	9
Coulde roodes pasture	14	2	14
The new close arable	14	0	0
Midlewood spring close arable	10	2	13
Peaseflat arable	7	2	17
Thistlehowme pasture	11	0	20
Normanhowme meadowe	9	2	24
Bridgeroyde meadowe	3	3	25
Stackclose pasture	22	1	15
Nun wood	33	0	0
Nun wood closes in the ocupation of Steven Jacksome	4	2	35
Sm: totalis the Demayne	343	0	16

The farmes

	Acars	Roodes	Perches
Robert Gaskone howse and groundes called Nun wood	11	0	26
Thomas Bridges howse and groundes called nun wood	11	1	0
Christofer Lacocke howse & nun wood closes pasture & arable	6	1	28
Harrison pighill	2	0	17
John Wilson howse and groundes	4	1	34
Thomas Beeston farme	14	1	9
Christofer Dunwell howse and crofte	0	2	0
One close called Viccar ing meadowe	3	0	13
The Litle ing meadowe	1	1	27
One other close pasture	6	1	0
Sm: totalis Dunwell	11	1	0
Sm: totalis	404	0	6

mented and numbered at 16, 32 and 48. Above is written: 'This Scale Conteyneth 48 perches', giving the scale of sixteen perches to one inch.

This is an interesting detailed map, but unfortunately some details are partially obscured with age and the thick colouration. Pontefract Castle, a gatehouse drawn symbolically in outline, is shown in the north-east; Carlton village in the south-east; part of Pontefract Park in the east and north, and part of the West Feilde in the south. Park paling is drawn along the boundary of the park, with one entrance gate only drawn on a road from Pontefract town. A number of buildings are shown on the roads leading into Pontefract, but as the map is of Tanshelf, Pontefract is not shown in detail and there is a tantalizing blank space in this section of the map. With the exception of parts of three open fields the land is divided into small enclosures in each of which is written the tenant's name, the field name and sometimes the land use, i.e. meadow, pasture or arable. Roads are drawn as before, either dotted lines for unfenced sections, or with continuous parallel lines in the enclosed fields and areas of settlement. There are no trees or indication of high ground.

The accompanying written survey is headed: 'A Survey of the Demanes of the Mannor of Tanshelf taken by Robert Saxton An̄: Dni: 1611'. There are thirty-seven tenants listed, together with the names of the closes each occupied, often followed by the word meadow, arable or pasture. Mr Savage held the greatest quantity of over forty acres, but many tenants had only one small field. Then follows a list of 'Cottages': seven men who had simply a house and garth, or garden; and finally the wastegrounds comprising Swine Lane, the Marlepit and Carlton Green Common. The whole survey covered a total of 186a 0r 0p. Again, Robert Saxton does not use dayworks, simply acres, roods and perches as measurements. The survey is written by Robert Saxton himself, but at the bottom is the signature 'W Currer Surveyor', who presumably either assisted in making the survey, or certified it as correct. Currer was Surveyor of Lands for the North Parts, an appointment of the Duchy of Lancaster, made in 1608.[16]

The manor of Tanshelf was part of the Honour of Pontefract and in the Duchy of Lancaster. No specific reason has been discovered for Robert Saxton's employment to map and survey the demesne lands; it was probably a general survey of the tenants' holdings there in 1611.

Thornhill, Yorkshire

There is evidence that Robert Saxton was employed by Sir George Savile in 1612 in connection with his estate at Thornhill. The survey is described on page 97, and a description need not be repeated here. It is perhaps worth noting that Robert Saxton worked at Thornhill in April 1612, the date of his additions to the survey. It might be presumed likely that had his father Christopher still been alive and well he would have been asked to bring the original survey of 1602 up to date; after all he was very familiar with the area. Here, however, Robert has taken over, supporting the theory that his father was no longer available. On the other hand, it is quite probable that Robert assisted his father in the original survey and was also familiar with the territory.

Esholt, Yorkshire

In that same year, 1612, Robert Saxton was employed to survey the Lordship of Esholt, which lies four and a half miles north of Bradford and ten and a half from Dunningley.

Three copies of the written survey have been traced, two are at the Yorkshire Archaeological Society,[17] and one in the Leeds City Archives.[18] Only one of the former is in Robert Saxton's hand. This covers one sheet of paper only and is headed: 'A Suruey of the Lōdship: of Eshesholde taken by Robert Saxton A⁰. Dm̄: 1612.' Land measurements are given in acres, roods and perches. Over half the survey of 343 acres is concerned with the demesne grounds, the remainder lists the farms with the names of the tenants, their holdings and acreage. The total area surveyed was 404a or 6p. The survey is reproduced on page 128 to illustrate both Robert Saxton's writing and a typical survey.

Following the dissolution of the Monasteries, Esholt Priory remained in the hands of the crown until 1547 when it was conveyed to Henry Thompson, one of Henry VIII's gentlemen at arms. The Letters Patent of this grant are quoted in full by Slater and list the lands attached to the Priory at Esholt at that time, many of which are identifiable in Robert Saxton's later survey.[19]

The Lordship of Esholt remained in the hands of the Thompson family until Frances Thompson, Henry's great great granddaughter, married Walter Calverley, son of Henry Calverley, who had owned the Calverley and Farsley lands surveyed by Robert Saxton in 1609. Their grandson sold Esholt to Robert Stansfield of Bradford, in whose archives one of the surveys survives.

Presumably the survey was made simply as an assessment of the holdings of the Thompsons in 1612.

Manningham, Yorkshire

Both a map and written survey by Robert Saxton of the township of Manningham survive in the Public Record Office.[20] Manningham lies one mile to the north-west of the centre of Bradford, and ten miles from Dunningley.

The map measures 58cm by 86cm, with the title in the top left corner, which reads: 'A plat of all the landes within the Towneship of Maninggam Wherein The old land freholde is Colored with red, the old land Coppyholde with grene The New land Coppyhold with yelow and the new land frehoulde lefte white; the old land on the Eastide of the towne being colored with grene: and red is meadow ʊ arable; the feilds old land liing on the weste side the towne is Arable and pasture: taken by me Robert Saxton Anͦ: Dm̄ 1613 . . .' The scale, in the bottom right corner, comprises a pair of open compass dividers surmounting a segmented rule measuring three inches, numbered at 20, 40 and 60. Above is written 'This scale conteineth 60 perches', giving a scale of twenty perches to one inch. In the bottom left corner is written, in an unknown hand: 'This platte was deliuered up by Robert Saxton Surveyor abovenamed to be inclosed the Eleventh daie of June instant 1613 according to his oath taken for the same purpose the Eight & twentith daie of May last past before vs, viz: Charles Clapham & Willm̄ Harrison by virtue of his highnes Commission to vs & others to that ende directed.'

The map is very detailed and comprehensive, showing the whole township of Manningham, and naming those adjoining. Roads, watercourses, buildings and fields are shown. The occupants, both freehold and copyhold tenants are named, together with some, but not all, land usage, pasture, meadow or arable. No trees or high ground are depicted. The map is coloured as noted in the title, but this has become faded with age. Although much divided, the old town fields are still distinguishable and named. Crossing the map in a north-westerly direction is 'The Kinges hye Strete', running from 'Bradforth' to Heaton Common, now known as White Abbey Road and Toller Lane. For further discussion of the contents of this map the reader is advised to consult *The Township of Manningham in the Seventeenth Century* by Wilfred Robertshaw, who also reproduces the map.[21]

The written survey of Manningham, transcribed in full by Robertshaw, is of some length. It contains details of occupiers, the names of the fields held and the acreage in acres, roods, dayworks and perches. It is interesting to note that Robert Saxton reverted to using dayworks on this occasion. Distinction is made between land held in the common fields, old land freehold, old land copyhold, and new land in both categories. The acreage of each man's land is totalled up, but as is often found, the additions are erratic. The total area surveyed was over 662 acres.

At the end of the survey is a note: 'The laines and hye waies in this township of Manningam are measured as the rest but the Contente not set downe, but easily may appear by the scale if nede require, by me Robert Saxton Juneior'. This is followed by another note, as on the map, about the survey being delivered by Robert Saxton according to his oath. The declaration that Robert Saxton took his oath on May 28th and delivered the map on June 11th, only fourteen days later, shows at what rapid speed he worked. If one takes into consideration time taken for travelling to and from Manningham, time taken to draw the map and write out the survey after the fieldwork was completed, and the time taken in gaining all the necessary information concerning occupants and tenure, it is probable that only about a week was available to survey the township.

Robert Saxton was commissioned to make the map and survey in connection with proceedings brought by Charles Tirell, relator on behalf of the crown, against Richard Wilkinson, William Northropp and Nicholas Crabtree, who claimed to hold certain 'new lands ... taken out of the wastes' since the year 30 Elizabeth (1587–8). The survey was to discover whether these were part of the manor of Bradford, or concealed from the crown.[22] The holdings of the three named above are all detailed in the survey and on the map, but in only one instance does Robert Saxton refer to the dispute; in the survey the list of William Northropp senior's holdings includes 'Magget inges in question betwixte Bradford and Manningham ... 6a 3r 6dw 3p'.

Liversedge, Yorkshire
Frank Peel in his book *Spen Valley Past and Present* recorded that Christopher Saxton surveyed part of Liversedge in 1608 (see page 119). Peel also wrote that Robert Saxton made a map of the common land of Liversedge in February 1614. Again this map appears to have been in Hanson's collection of archives which cannot now be traced, and the information below is taken direct from Peel's book. Liversedge lies five and a half miles southwest of Dewsbury, between Halifax and Batley, in the West Riding of Yorkshire.

Sir Phillip Carey held the manor of Liversedge in 1614 and in attempting to increase his own possessions proposed to the thirty-two other freeholders that they enclose the 456 acres of the town's common land and divide it amongst themselves in proportion to the parish 'lay' or assessment they respectively paid. By this means Sir Phillip stood to gain about half of the land. Peel says: 'When Mr Robert Saxton, who like his father was an "experte surveyor," showed them ... how the spoil could be fairly divided, according to Sir Phillip Carey's notions of fairness, they unanimously agreed to the scheme, and the matter was proceeded with. A rough ground plan has been handed down, but as it is not shown upon it how the land is bounded on any of its sides, it is of little practical use, and our ideas respecting it can only be of the crudest sort.'

Peel then goes on to describe how the land was divided as shown on the map. There was also a table listing the freeholders and how much land they received for their lay of, however much it was. The land was divided into three kinds – best, worse and worst, and all had his proportion of each. John Hanson, who was a freeholder, received 14a 1r 13p 'for his laye of 1s 9d', the third largest portion. He notes on the table, which was evidently his, '1614. – Memorandum that Mr. Robert Saxton, the surveyor, Richard Rayner and others, did upon the 19th, 20th, 21st, and 24th days of February divide the moore of Liversedge amongst the freeholders respectively for every man's rate according to a laye by them or some of them agreed upon. And my first lot in the first division was the twelvth in order, and in that division I had three acres, three roods, and eight pearches. My second lot was the ninth in order and layeth about Cockell, and by south the Highgate, I had in that divition, five acres, one rood, and nine pearches. And my third and last Doole was made upon the 24th day of February and I was in No. the 7th, and I had in that division five acres, one rood, and fourteen pearches of the flatt called Walestone croft flatt, and a parcell about my close at the hyerood.'

This extract is interesting in that it shows in what manner each freeholder received his portion. The first day the best land was distributed, the second the second quality, the third the worst. It almost sounds as if they drew numbers out of a hat in that it was worthy of note in which order they drew their lots. Also noteworthy is the complete indifference to any claims of other than the freeholders to these common lands, but that was usual practice at the time.

It is unfortunate that none of the original papers concerning the enclosure of the common at Liversedge can be traced. They cannot be found at the Bodleian Library, and as Peel gives no specific reference, merely saying that he discovered some old papers referring to the Manor, 'which seem to have remained undisturbed for more than three centuries in the Bodleian Library, Oxford, where they were desposited by an antiquary named John Hanson,' there the matter must rest, at least until the papers are rediscovered.

Hatfield, Yorkshire

In 1615 Robert Saxton was employed by the crown in connection with proceedings relating to waste land at Hatfield in the south-eastern part of the West Riding of Yorkshire, nine miles north-east of Doncaster, and some thirty miles from his home at Dunningley.

A map drawn by Robert Saxton and documents relating to the proceedings survive at the Public Record Office.[24] Unfortunately the map is badly damaged, with the bottom left quarter, the top right corner, the bottom edge and part of the title missing. What remains measures 83cm square, but when complete must have measured about 104cm by 83cm. The title, in the top left corner, reads: 'A plat of ... groundes and wastes within the ... of Hatfeilde in the Contriuersye betwixte Sir Frãcis Bacon Knight his plan: Aturney Generall playntyf and William Bennitland Richard Brewer and otheres defendantes wherein Heynes planes, Durkenesse planes Tockledge, and Northtoftes, are Colored with Grene, the Comõn and pasture betwixte Heynes and Ducklingsike and ducklingLedge which dooth ioyne boothe together is Colored with yelowe, Inglemores is Colored with Orredge, and the Confines left white made by Robert Saxton Aº. Dni 1615.' Added at a later date beneath the above, but within the cartouche, is 'Midlinges which the tenits of Thorne doath howlde by rente is colored with red. Jonath: Kaye.' The scale rule is missing.

The area shown on the surviving portion of the map includes 'Inglemores The Lo. of Hatfeildes soyle' in the north-west, basically Thorne Waste, and 'Croole in Lincolneshire' in the east. In the extreme west is drawn an isolated church which is simply named 'Thorne'. From Thorne running due east is a road, drawn as double dotted lines, which follows the route later taken by the Stainforth and Keadby canal. Tockledge is shown in the south, with the river Don ('Dun flu') dividing Yorkshire and Lincolnshire.

The original colouring has faded, but is generally distinguishable, with a typical red border. The map as a whole, covering a large area as it does, appears rather plain and empty, with no details of fields or other boundary features. Only three buildings, Thorne and Crowle churches and Tockledge Lodge, are shown. There are no trees and no high ground. Although there is no scale shown this map is on a much smaller scale than most of Robert Saxton's other maps.

Hatfield Chace was crown land at this date and controversy had arisen over the waste lands in the Chace. A Special Commission of Enquiry was set up and Robert Saxton engaged to map the moors, commons and waste. There were lengthy disputes over how the commission was to be executed, but the case was eventually heard in August 1616 when Sir Francis Bacon, the attorney general appeared as plaintiff and William Benetland, Richard Brewer, Robert Stevenson and Robert Palmer appeared as defendants, named on behalf of the tenants. The Commission met in the public Court House at Hatfield on August 27th and a jury sworn in before the Sheriff of the county, Nicholas Hunt and his bailiff of the Hundred of Stafford. They did then and there 'open and examine the map or plat anexed to the said Commission touching the names feilds and bounders of the wast comons and mores . . .' Witnesses were examined and the following day the commissioners, jurors and witnesses 'did view perambulate and survey the said comons moores and wastes all day then adjourned to Hatfield'; it must have been quite a gathering! The next day, August 29th, more evidence was taken and the jurors gave their verdict, that 'this plot map and description of the said Moores Comons and waste ... all the names of the bounders in the plott or mapp . . . is true as did appear unto us by viewing of the ground'.

The map is particularly interesting in that it shows the old course of the river Don and the area before it was drained by Vermuyden in the reign of Charles I. The area was extremely prone to flooding and was of little use except as a royal hunting ground. The absence of features on the map would indicate the true state of the territory at that time: no highland, fields or enclosures, and sparse habitation.

The discrepancy between the date of the map and of its examination can more easily be understood if it is remembered that the year commenced on March 25th, and if Robert Saxton had prepared his map in the early spring it would be dated the previous year.

Ledston, Yorkshire
A survey of the Lordship of Ledston, four miles north of Pontefract and nine miles east of Dunningley, survives in Leeds City Archives.[25] It is on a long paper roll, and is in Robert Saxton's writing. The title reads: 'A Survey of the Loship of Ledston taken by Robert Saxton Aº: Dm: 1616.' Measurements are described in acres, roods and perches, with sum totals of each tenant's holdings.

The survey commences by describing roughly 250 acres of demesne lands 'in the Lo: handes' which included 'the Scite of the hall with the courte Orchard & Garden' covering almost six acres. Next follows the 'Demanes in the teñor of Mrs Withu' of 158 acres, a list of the other tenants and a detailed summary of field land, enclosed ground, stinted pasture, common and woodland. It is a very detailed and interesting survey to study, throwing light on the agrarian system of the time. The tenants each held a portion of the common fields and one or more beastgate in the 'firth' or the 'more'. All fields are named. The mathematical additions however, are somewhat erratic.

The manor of Ledston was owned by the Witham family who lived there for three generations. William Witham of Ledston Hall died in 1593, supposedly by witchcraft, and his son, Henry Witham, sold the estate to Sir Thomas Wentworth, afterwards Earl of Strafford. Negotiations for the purchase of Ledston were protracted and it is not clear whether Robert Saxton was employed by Henry Witham or Sir Thomas Wentworth to survey the estate in 1616, although the survey must have been made in connection with the sale. In the same collection is another survey of Ledston, dated 1617, made for Wentworth, and this is accompanied by a map of contemporary date, neither of which can be attributed to Robert Saxton. As these were made for Wentworth, it seems logical to expect that Robert Saxton's survey was made for Witham.

The 'Mrs Withu' named in the survey was Henry Witham's mother who lived until 1619. Sir Thomas Wentworth lived at Ledston for many years including during the period when he was President of the Council of the North in the reign of Charles I.

Sutton in Craven, Yorkshire
A further written survey in the Leeds City Archives was compiled by Robert Saxton for a Mr Copley in 1616.[26] It is of Mr Copley's land in Sutton, in the parish of Kildwick, five miles north-west of Keighley in the West Riding of Yorkshire.

It is a detailed survey, written by Robert Saxton himself, on a single sheet of paper, but it would appear that this sheet comprises only half the survey and that there was originally a second sheet, now lost. This is apparent from the existence of a copy of Robert Saxton's survey in the same collection, which is twice as long. The original is headed: 'A Suruey of M^r: Copley his land in Sutton taken by Robert Saxton and John Haigh Ann°; Dni: 1616.' It details the holdings of five identified tenants, naming fields and closes held and giving land measurements in acres, roods and perches. Distinction is made between enclosed and unenclosed land. In the left-hand margin, against most of the lands listed, appear the letters A or B, or occasionally AB, or C. No significance has been discovered for these letters. The mathematical additions are worthy of note in that they are accurate! There are some later monetary figures added in the right-hand margin, probably the annual value. There is no record of a land surveyor named John Haigh, possibly he was an employee of Mr Copley's in another capacity and delegated to assist Robert Saxton. A John Haighe is named as a witness to Thomas Saxton's Will in 1608, which may imply a closer association with the Saxton family.

The second survey, which has the same title, is written in an unknown hand. It is not an exact copy and has been slightly altered; whether this was intentional or merely an inaccurate copyist is not clear. The survey is twice as long as the one in Robert Saxton's hand, suggesting that the latter is incomplete.

The Copleys were an ancient family, Lords of the Manor of Batley. Edward Copley of Batley Hall died in 1616 and administration of his Will was granted to his son Alverey Copley on April 12th of that year. It was a notable year for Alverey, for not only did he inherit, but he married Elizabeth, daughter of Sir John Savile of Howley Hall.[27] The marriage settlement details certain lands including the manor of Sutton-in-Craven and appurtenances in Sutton, and it was presumably for this that Robert Saxton was employed to make the survey.[28]

Oakworth, Yorkshire
In 1617 Robert Saxton was again called upon to map land in a case before the Duchy of Lancaster Court at Westminster. It concerned the disputed boundary between the manors of Colne, in Lancashire, and Oakworth, in Yorkshire, which was in effect a dispute over the location of the boundary between the two counties, and thus of considerable importance. Oakworth lies three miles south-west of Keighley, in the West Riding of Yorkshire.

The map which Robert Saxton prepared in evidence is in the Public Record Office and measures 50½cm by 66½cm.[29] The title, in the top left corner, reads: 'A plat of the Common In Question betwixte the Inhabitāts of Wycoller and Okeworth wherein the Comon in Question is Colored with Grene ϛ the Confines lefte white made by Robert Saxton A°n:

Dni:.1617.' The scale, in the bottom left corner, comprises a pair of open compass dividers surmounting a segmented rule measuring four inches, marked at 20, 40, 60 and 80. Above is the statement: 'This Scale Conteyneth – 80 – perches', giving a scale of twenty perches to one inch.

The area shown is bounded to the north and west by parts of Wycoller pasture, to the south by parts of Stanbury Common and Inclosures and to the north-east by part of Keighley Common. Roads, watercourses and hills are shown and named, the latter being landmarks referred to in the depositions. There are no trees shown. The colouring is rather crudely applied, and, apart from the roads in brown and the streams in blue, is a thick deep yellow. There is a photographic copy of the map in Bradford Public Library.[30]

The manor of Oakworth, or Okeworth as it was then called, belonged to the Copleys of Batley until 1614 when Alverey Copley conveyed to John Midgley, Arthur Rawson, Christopher Mitchell and Robert Heaton 'all the commons mores and wast lands commonly called Okeworth Moor in Okeworth, Scholes and Deanfield and elsewhere in the Lordship of Okeworth, one acre of the said common abutting on a place called Wolfestones excepted', for the sum of £180. The four purchasers named were, in fact, trustees for the thirty freeholders of Oakworth who proceeded to divide the common between them. However, the western boundary of the common was disputed by the owners of the adjoining manor of Colne. The Oakworth freeholders claimed that the natural watershed was the boundary, whereas the owners of Colne claimed it to be below the watershed on the Yorkshire side. The manor of Colne was part of the Duchy of Lancaster and as such, James I was Lord of the Manor. The ensuing legal wrangle then was between no lesser person than the King and the freeholders of Oakworth, and it is perhaps surprising to what lengths the court went to seek evidence and a just settlement, but there is no doubt the verdict was predictable in finding for the King.

No less than four commissions were set up during 1617 and 1618, but the one that involved Robert Saxton is dated December 3rd, 1618.[31] Ten commissioners, including Robert Saxton, were appointed by the Duchy Court and the case was between Nicholas Cunliffe, plaintiff, and Henry Pighell and three others as trustees for the freeholders of Oakworth, defendants. The commissioners were instructed that 'you or any twoe or more of yor at tyme and place convenient to yor to be lymited and appointed to call before you by vertue hereof the said pties Plator and Defts and all such other psons as yo^u shall thinke meete to inquire of the said matter aswell by viewe oathe evidence pambulacon examacon of witnesses as by all other good waies ... And o^r further will and pleasure is that you or any twoe or more of yo^u do take a pfecte viewe and survey of the lands in question and the Boundaries thereof. And therupon to make a pfecte plotte of the same lands wth the Boundaries Meares Markes and other noted places thereof according as they are claymed by either ptie and as by them shalbe desired.' It was endorsed by John Blakey, Robert Saxton and Robert Hey as acting commissioners. Depositions of witnesses were taken on January 14th, 1618, before five of the commissioners including Robert Saxton, described as 'gent', and these were also endorsed by the three named above.

There is an obvious discrepancy in dates; the map, being dated 1617, and the commissioners' (which included Robert Saxton) instructions, being dated December 1618. The most likely answer is that Robert Saxton drew the map for an enquiry in 1617.

The ultimate verdict of the court, that the boundary lay well below the watershed, has had far reaching repercussions. Watersheddles Reservoir built on the Lancashire side of the boundary, but on the eastern side of the watershed, by Keighley Local Board of Health in 1869, involved the ratepayers of Keighley in paying rates on the reservoir to both local and county authorities in Lancashire.[32]

Wensleydale, Yorkshire

T. S. Willan wrote a paper entitled *Three Seventeenth Century Surveys* which included details of a survey of the manor of Wensleydale in the North Riding of Yorkshire.[33] In 1616 the manor was divided into three and sold, one third being purchased by George and Thomas Cole, one third by John Goodman and his son John, who sold it to George Cole, and the final third by Mary Colby who leased her portion to George Cole. To settle the problem of dividing the manor into three equal parts a Special Commission of Enquiry was set up by the Court of the Exchequer, and eight commissioners were appointed. Robert Saxton was engaged to survey, measure and set out the common belonging to the manor, in April 1618.

Robert Saxton submitted his deposition at Askrigg and it is perhaps worth quoting in full as being the only evidence now remaining of his work.[34]

'Robert Saxton of Dunningley in the countie of Yorke gent of the age of xxxiij yeres or thereabouts sworn and examined deposeth and saith as follows ... saith that he was sent for to ... and comeing thither he was intreated to survey measure and sett oute Three thousand three Acres of comon belonging to the mannor or Lōpp of Wensladale as neare as hee this depot could, to a plat or mapp deludd him formerly made by one Sammarll Pearce as he was informed whereupon he the dpont went to the said Comon with the plaitis and assent of George Cole Es. And Marye Colby wid^w and ... there vizt. the vijth lxth xlth and twelft daies of this instant April in measureinge and surveyinge the same and accordinge to the said platt or mapp deludd him And ... Thinketh that in his conscience rather to be lesse than more than threethousand three Acres And hath butted and bounded yt as is sett downe in the peticon whereunto this depos is sett down. But the depo^t knoweth not of his own knowledge whether the names of the buttalles or bounderes be true but only sett them downe as he was informed by the compani which went with him and were with him being tenants of the said Lordshipp.'

Once again it seems that quite a gathering of men accompanied the surveyor on his perambulation. It is apparent from this that Robert Saxton did not in fact draw an original map of the manor, but rather was given and used a map previously drawn by Samuel Pierse. Pierse had been employed to survey the manor of Wensleydale in 1614 and it is interesting to note that he, like Christopher Saxton, used dayworks in his land measurements. Pierse came from Maidstone in Kent, and the daywork was more commonly used in that county than elsewhere.

One other very interesting point emerges from Robert Saxton's deposition. He states his age and status. This is the only evidence of his age from which can be calculated his approximate date of birth as 1585, 'or thereabouts', and from the fact that he was Christopher's eldest son, the approximate possible date of Christopher's marriage can be roughly deduced.

The map in question, however, cannot be traced, and must now be considered lost.

Gomersal, Yorkshire

At the time of writing the last known work of Robert Saxton was in 1619 when he roughly mapped and surveyed land in Gomersal, although no doubt, as he lived for another seven years, other later examples of his work may be discovered.

Gomersal, some five miles west of Dunningley, is in the parish of Birstal in the West Riding of Yorkshire. Both the map and survey are now at the Yorkshire Archaeological Society.[35] The map has no title, but is endorsed, in an unknown hand, 'A Survey of that pt of land belonging to Elyzabeth Gombsall daoghter to John Gombsall of lyttle gombsall layt deceased', and, beneath, 'by Saxton'. There are other endorsements noting that statute measure was used and some calculations. The map itself is not drawn by Robert Saxton and is in fact more of a rough sketch map than is usually associated with his draughtsmanship. It measures 39½cm by 31cm, has no scale or margin, and the cardinal points are written in script, almost carelessly, on the map face. Eighteen fields are shown, some of which are detached; all are named, with measurements given in acres, roods and perches. Only one house is drawn, but the name is obliterated by a fold in the paper. There is nothing on the map to indicate where the fields are situated in Gomersal.

The accompanying survey is written by Robert Saxton in his own hand, on a single sheet of paper which is torn. The title reads: 'A Suruey of all the Landes in Gumbersall which did belong to John Gumbersall late disseased taken by Robert Saxton Anº: Dm̄: 1619.' It is endorsed, in the same hand as the endorsement on the map, 'Saxton Survey of yᵉ whole lands 1619'. The survey is set out in three sections, firstly '(The) South (part)', which lists a number of fields totalling 60a 3r 29p; secondly, 'The North part', listing fields totalling 63a 3r 9p; and finally, an unnamed section listing fields totalling 27 acres. The survey concludes with the 'Smᵃ totalis – 151. 2. 38.' The fields named in 'The South part' correspond with those drawn on the accompanying map, which may suggest that another map or maps were originally made of the remaining land.

Elizabeth Gomersal, daughter and heir of John Gomersal, married Francis, second son of John Popeley of Woolley Moor House, and their daughter Grace married Sir Thomas Wentworth of Bretton where the documents remained until recent years.

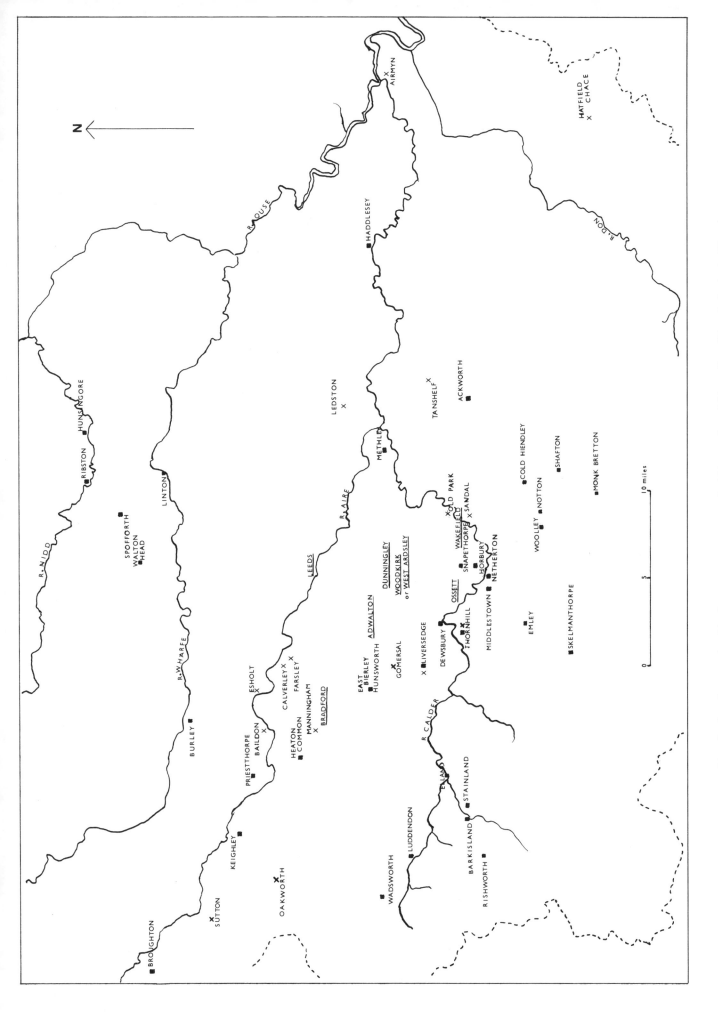

Map showing places mentioned in the text and places surveyed within a radius of thirty miles of Dunningley in the West Riding of Yorkshire. Places marked by a square were surveyed by Christopher Saxton, those marked by a cross were surveyed by Robert Saxton.

PLATE 19

Conclusion

The appearance in England of a cartographer of national repute was not altogether surprising during the latter years of the sixteenth century. Map-making of a sophisticated order was firmly established abroad, especially in the Low Countries, and already some rather inadequate efforts had been made to record British geographical information. The opulence of the Elizabethan court, however, combined with a latent necessity for recording our geographical environment, almost inevitably produced the right man at the right time. Christopher Saxton executed the maps for which demand had been growing over several years, while the investigative and adventurous spirit of the age encouraged him.

There remain, nevertheless, significant mysteries surrounding the person. His first apparent achievement was his greatest, namely the completion of a national survey and its representation in maps. Nothing concrete survives from earlier years. Yet Saxton's emergence from an obscure Yorkshire background in aspiration towards national eminence must surely imply some earlier activity upon which we remain ignorant, for the time being at least. His subsequent preoccupation with localised estate surveying is not perhaps altogether strange: after all, he had successfully mapped the country as a whole. Yet even here there remain perplexing situations. Why did he not, for example, as John Speed was shortly to do, seek to popularise his work and produce another atlas on a yet more grandiose scale? Instead he devoted his time to the professional and probably mundane business of estate survey. Not much of this work has come to light, however, and many surveys which have been discovered relate to Yorkshire, implying a gradual retreat to his native heath. More will no doubt appear in the course of time, and this book has listed several hitherto unreported; the comparative rarity of Saxton surveys, however, cannot but be regarded as somewhat surprising in view of his established reputation, and his withdrawal to the north is more than a little odd for somebody who had achieved fame in the metropolis. There are, nevertheless, some very positive links. The probable association with John Rudd, from whom he must have learned his trade, and the inculcation of skill and interest in his son Robert, both provide a ring of authenticity; suggesting that behind Christopher Saxton there lies a complex story which, despite our considerable knowledge, we have perhaps only begun to unravel in human terms.

His influence cannot be disputed. Not only did he begin the development of reputable English map-making, but he established its direction for many years to come. Individuals such as Norden, Speed and Drayton, to name but a few, all embarked on similar atlas-making ventures. And the employment of the county as a mapping framework survived even into the age of precision which was formally recognised by the establishment of the Ordnance Survey. Investigation into the geographical as opposed to carto-bibliographical significance of his maps has not proceeded far, and there remains much scope for research in this field. Such work will undoubtedly cast further light on the man and his times, and will serve to clarify some points which remain perplexing and obscure.

There seems to be no positively established cartographic activity in the Saxton family after Robert. Perhaps brief mention should be included, however, of a survey in the Leeds City Archives (HE/1) made by an Anthony Saxton in 1646. The survey is of Tadcaster in the West Riding of Yorkshire and is contained in a large paper book. On the cover is floridly written 'Anth. Saxton Booke 1646'. It is interesting to discover a later surveyor named Saxton working in Yorkshire, but no connection with Christopher has been traced.

Appendix 1

EXTRACTS FROM THE LAY SUBSIDIES 1547–1621

OSSETT
1545	wife of Saxton	in goods	£6	12d
	Robert Saxton	,,	£2	2d
	John Saxton	,,	£1	1d
1567	Robert Saxton	,,	£4	3s 4d
1571	Robert Saxton	,,	£3	3s
1585	Robert Saxton	,,	£1.6.8.	3s 7d

WEST ARDSLEY
1567	Thomas Saxton	,,	£4	3s 4d
1571	Thomas Saxton	,,	£3	3s
1585	Thomas Saxton	,,	—	
1588	Thomas Saxton	,,	£3	5s
1597	Thomas Saxton	,,	£3	8s
1598	Thomas Saxton	,,	£3	8s
1599	Thomas Saxton	,,	£3	8s
1605–6	Christopher Saxton	,,	£3	5s
1620–1	Robert Saxton	,,	£3	3s

Appendix 2

PARISH REGISTER TRANSCRIPTS – WOODKIRK als WEST ARDSLEY, YORKSHIRE

- 1600 Thomas Saxton was buryed the fourth(?) day of October.
- 1601 Thomas the son of Christopher Saxton was baptzd the xxij day of November.
- 1605 Christ: Saxton & Jasper Lynley signed as Churchwardens.
- 1608 Williã: son of Robert Saxton bapt Novemb 6.
- 1623 Thomas Saxtonn and Mary Thompsonn married the xiii day of November.
- 1623 Christopher Saxton marryed the iii day of January to Prudence Hartley.

Appendix 3

EXTRACTS FROM THE FEET OF FINES 1581–1622
(Yorkshire Archaeological Society Record Series Vol.V (1888))

1581 Plaintiffs: John Saxton and Robert Saxton.
Deforciants: Thomas Saxton and Marjery his wife.
Messuage with lands in Woodkirke, Dunynglawe, Tynglawe and Westardslawe.

1609 Plaintiff: William Saxton.
Deforciants: Robert Saxton and Anne his wife.
Messuage and lands in Horbury.

1614 Quer. Robert Saxton and Roger Casson.
Deforciants: Nich. Nalson and Alice his wife and Thomas Nalson.
Messuage, 2 cottages and lands in Altofts.

1616 Quer. Gregory Armytage, gent.
Deforciants: Robert Saxton, and Anne his wife and Anthony Glover and Agnes his wife.
Messuage and lands in Tingley, Duningley, Topclyf and West Ardsley.

1616 Quer. William Brooke and Thomas Hirste.
Deforciants: Robert Saxton gent., and Anne his wife and Thomas Wheatley gent.
Messuage and lands in Dalton and moiety of the fairs and feasts of Adwalton and a moiety of the tolls, stallages, picages and bridge tolls within the manor of Adwalton and a moiety of the Court of Piepowder, liberties and franchises of Adwalton.

1616–17 Quer. Robert Saxton.
Deforciants: Robert Greenwood gent. and Jane his wife.
Messuage, cottage and lands in Dunningley.

1617   Quer. Anthony Glover.
Deforciants: Leonard Waddington and Christian his wife and Robert Saxton and Anne his wife.
Messuage and lands in Woodkirke and West Ardislaw.

1619   Quer. John Nettleton.
Deforciants: Robert Saxton and Anne his wife.
Lands in West Ardsley and Dunnyngley.

1622   Quer. Gregory Armitage gent.
Deforciants: Nicholas Linley and Dorothy his wife, Robert Linley and Robert Saxton and Anne his wife.
Messuage and lands in West Ardsley.

Appendix 4

THOMAS SAXTON'S WILL
(Borthwick Institute of Historical Research, York. Prob.Reg.31 Fol.337v, 338)

In the name of god Amen the 14th day of January in the yeare of o^r lord god 1608 I Thomas Saxton of Duninglaw in the County of yorke yeoman being sicke in bodye but whole in mynd and of good and pfect remembrance thankes be given unto god therefore do make this my last will and testament in manner and forme following ffirst I give my soule unto Almightie god and my body to be buryed in the pish church or churchyard of West Ardley alias woodchurch in the Countye aforesaid and the church to have the dutyes due for the same Itm I give and bequeath to my brother xpofer Saxton and to my sister his wife either of them xij^d Itm I give to my sister Alice Hobson xij^d Itm I give to my brother in lawe Peter Page and to my sister Margrett his wife either of them xij^d Itm I give to Robt Saxton the eldest sonne of xpofer Saxton my said brother xij^d Itm I give to xpofer Saxton the second sonne of my said brother xij^d Itm I give to Thomas Saxton the third sonne of my said brother one stand bedd standing in my brothers plor Item I give to the daughters of my said brother xpofer Saxton evry one of them xij^d Itm I give to Mathewe Page the sonne of the aforesaid Peter Page my brother in lawe xij^d Itm I give to Mary Hobson xij^d Itm I give to my servant Elizabeth Blades xij^d Itm all the rest of my goods my debts being paid my funerall expenses being discharged I give & bequeath to Alice my wife Itm it is my will that the said Alice my wife shall during her life quietly and without lett molestacon or hinderance have, hold, enjoy, occupie and possesse all my land according to the true Intent and meaning of one deed of ffeossmt passed unto certaine in trust to the use and behoofe of the said Alice my wife after the day of our mariage as on the said deed it doth and may further appeare And also I do further give and bequeath unto my said wife and to her heires and assignes for every one close pcell of my said land usually called and knowne by the name of the longlands And it is also my will that after the decease of my said wife all my said land (the aforesaid close called the longlands onely excepted) shall discend and remayne unto the two youngest sonnes of xpofer Saxton my said brother that is to say to xpofer Saxton & Thomas Saxton and unto there heires and assignes for ever Prouided always that xpofer Saxton my said brother and his heires and asignes whatsoever shall in consideracon hereof suffer my said wife during her life quietlie without anye manner lett contradicon molestacon or hinderance to have hold occupie and enjoye all that mansion or dwelling house wherein I nowe dwell with all other the edifices and buildings thereunto belonging and also the croft and garden thereunto appteyning togeather with that way for cart and carriage which I nowe use into the west royd But if either my said brother or any his heires or assignes whatsoever shall refuse so to do or shall contradict or any waies molest or trouble my said wife in or about the premisses or in or about any pt or prcell thereof then it is my will that the-foresaid bequest touching my said land made unto the said xpofer Saxton and Thomas Saxton the younger sonnes of xpofer Saxton my said brother and there heires and assignes shall be of none effect as if it had not bene made at all unto them for in such case I give and bequeath all my said landes unto my said wife and to her heires and assignes for ever Itm of this my last will and testament I make Alice my wife the full and sole executrix And supervisor hereof I make my loving and trustie freindes Mr John Brooke clerke and parson of Emley and John Brooke of Adderton yeoman hopeing that as my trust is in them they will see the same truly executed and performed In witnes whereof I have hereunto sett my hand and seale the day and yeare first above written Witnesses hereof are Jasper Lynley John Haighe Anthony Glover. Proved on the 20th day of June 1610 by Alice Saxton widow the relict and sole Executrix.

Appendix 5

ROBERT SAXTON'S WILL
(Borthwick Institute of Historical Research, York. Prob.Reg.39 Fol.8)

In the name of god amen the five and twentith daye of September Anno Dni 1626 I Rob^{ti} Saxton in the Parish of Wood Church in the Countie of yorke yeoman beeing weake in bodie but of good and perfect memorie I praise god doe here make my last will & testament in manner & forme followeing ffirst I bequeath my Soule unto the handes of Almightie god & Jesus Christ my redeemer most humblie beseeching him to pardon all my great and maniefould sinnes bessechinge him to worke in me a holie faith & sound repentance persuading myself that in the blood of Jesus Christ I shall have all my sinns washed away And my bodie to bee buried in the Church of Woodchurch And for my worlie Estate thus I dispose of it thus to give unto his wife & sonne all my Estate my landes and my liveings my debts & funerall discharged whom I doe make my Executors of this my last will and Testament And Mr Anthonie Nutter minister, and John Reyner In witness whereof I have sett to my hand to this present writing Witness Henrie Gascoigne Theis legacies are excepted Imprinis to Christopher Saxton xl^s To Thomas Saxton and Marie xx^s to everie one of Roger Casson Children ij^s vj^d To Thomas Nalson my Sword to his daughter my god daughter Marie xij^s to my Sister Grace five shillinges and to my Sister Elizabeth x^s to my Cosen Marie Hobson v^s to everie one of her children ij^s vj^d to Robte Goltwhaite daughter my god daughter Elizabeth two shillinges and sixe pence, to my god sonne Willm Chadwicke j^s to Robte Casson ij^s vj^d. Proved on the 4th December 1626 by Christopher Saxton of West Ardsley clothier the brother of the deceased for the use and benefit of William Saxton the son one of the Executors then a minor.

Appendix 6

BURGHLEY ATLAS
(British Library. Royal Manuscripts. 18.D.III)

Bound up with Saxton's plates are eighteen manuscript maps, and two printed ones (Nos.48 and 57). Occasional lists and returns are also included. Column (i) locates each map in the atlas, and an asterisk indicates that the plate pertains to Saxton. Column (ii) gives the subject of the map or page, and in column (iii) are shown the engraved dates, where they are available. Information appended to various maps is explained more fully in column (iv). Details have frequently been added by Burghley to the maps themselves, but are instanced only where they are of particular interest.

(i)	(ii)	(iii)	(iv)
1.*	England	1579	*Accompanied by:* (a) *brief historical and topographical notes by Burghley;* (b) *number and names of parishes within the city, liberty and suburbs of London:* (c) *number of parsonages and vicarages extracted from the Queen's Records of First Fruits and Tenths in the Exchequer;* (d) 'Postes from London', *(with Burghley's computation of their yearly cost) to* (i) *Holyhead,* (ii) *Tavistock, and* (iii) *Bristol;* (e) *the ways and distances from various cities and towns to London.*
2.*	Cornwall	1576	*Accompanied by names of Justices.*
3.	Coasts of Devon and Dorset from Dartmouth to Weymouth, with a description of Ottermouth Haven		*Ms. Coloured.*
4.*	Devonshire	1575	*Accompanied by:* (a) *Justices;* (b) 'A Breef Note of ye places of Descent in ye County of Devon y^t are most daungerous and require greatest regard and assistaunce'; 'A Note what powder and match was appointed to be kept in store in every corporate towne'; *and* 'A Note of hir ma^{ties} Store of Ordonaunce, powder and match, lead, etc., remayning in ye L. Lietenauntes, etc. hands'. *All these notes occur in the map's margin;* (c) *division of the county for military purposes with the names of officers;* (d) 'Stages for Postes to be laid between London and Plimouth, 1595'.
5.*	Dorset	1575	*Accompanied by:* (a) *Justices;* (b) 'Dangerous places for landing of men in the county', *in lower margin.*
6.	Falmouth Haven		*Ms. Large coloured plan.*

143

7.	Isle of Wight		*Ms. Scale approximately 20 millimetres to one mile.*
8.*	Hampshire	1575	*Accompanied by:* (a) *Justices, to some of whose names Burghley has added particulars:* (b) 'A Note of the 7 deuisions in the countie Southampton, and what hundreds be within euerie diuision'; (c) 'The Companies and Numbers appoynted for y*e* reliefe of y*e* Isle of Wight', *with names of the captains.*
9.	North-eastern corner of Kent		*Ms. Endorsed by Burghley* 'Ryver of Stour betwixt Sandwych and Fordwych'. *With account of* 'Milton, in Saxon Midleton' (*Milton-by-Sittingbourne*) *in the time of King Alfred and Edward the Confessor, and a later note of post-stages* 'appointed to be layd betweene the court and Portesmouth, 1595'.
10.*	Kent, Sussex, Surrey, Middlesex	1575	*Accompanied by:* (a) *Justices for Kent, Sussex and Surrey;* (b) 'The Governoures of the liberties' *of London, and* 'the Justices of Peace in Midle(sex and) Shreves there without the libert(ies)'; (c) *Names of residents in Sussex, with their seats, in alphabetical order of the latter;* (d) *Justices for Middlesex.*
11.*	Somerset	1575	*Accompanied by Justices.*
12.*	Wiltshire	1576	*Accompanied by Justices.*
13.*	Oxfordshire, Buckinghamshire, Berkshire	1574	*Accompanied by Justices.*
14.	Berkshire		*Ms. Pen and ink. Scale approx. 25 millimetres to a mile.*
15.*	Hertfordshire	1577	*Accompanied by Justices.*
16.*	Essex	1576	*Accompanied by:* (a) *Justices;* (b) 'The title of the lawles Courte in the Honnour of Rayghle' *in 18 Latin rhyming verses, beginning:* 'Curia de Domino Rege, dicta sine lege', *and ending* 'Tenta fuit proximo die Mercurii post festum Michaelis A° dominae Reginae 30 (1588)'; (c) *names of the* 'xi walkes' *and walkers of* 'the forest' (*of Essex*). *On the map itself, at the mouth of the River Blackwater, Burghley has written* 'Heyghfeld fayre and fatt, Barndon park better than that, Coppledon beares a Crown, Copthall best of all'. *The last-named is probably Sir Thomas Heneage's seat near Epping.*
17.*	Suffolk	1575	*Accompanied by Justices.*
18.*	Norfolk	1574	*Accompanied by Justices.*
19.*	Northamptonshire, Bedfordshire, Cambridgeshire, Huntingdonshire, Rutland	1576	*Accompanied by:* (a) *Justices;* (b) *short pedigrees of Colvil of Newton, Pinchbeck of Pickworth, and Bendish of Leverington, with lists of other names.*
20.	Map of the same district as that shown on Map 19		*Ms. Pen and ink. Contains names of hundreds in Northamptonshire.*
21.	A LIST		*A list of* 'Knights of the Bathe' *made at the coronation of James I, July 25th, 1603. Followed by the names of ten other Knights* (*Bachelor*) 'made at Whitehall the 22 daie of Julie, 1603'.

22.*	Staffordshire	1577	*Accompanied by Justices.*
23.*	Warwickshire, Leicestershire	1576	*Accompanied by Justices.*
24.*	Lincolnshire, Nottinghamshire	1576	*Accompanied by:* (a) 'The names of all the townes within every sessions of the said countie (of Lincoln)'; (b) 'A briefe of all the men trayned and vntrayned furnished of everie sessions (in Lincoln)'; (c) *Justices of Lincolnshire, for the three divisions of Kesteven, Lindsey and Holland, and of Nottinghamshire.*
25.*	Derbyshire	1577	*Accompanied by Justices.*
26.*	Worcestershire	1577	*Accompanied by Justices.*
27.	Worcestershire		*Ms. Coloured. Burghley has written in the margin a list of residents, and inserted a few names on the map itself.*
28.*	Yorkshire	1577	*Accompanied by:* (a) *a plan of the Humber (map 29);* (b) *a plan of Scarborough Castle (map 30);* (c) *a list of parks and castles in Yorkshire and Worcestershire, and of* 'Hamlets in the Northridding belonging to Mydlam' (*Middleham*).
29.	'river of Humber and of the Sea and Seacoost from Hull to Skarburgh'		*Ms. Coloured plan. Includes particulars of tides, and list of* 'Havens and Crickes on ye Northsyde of Humbre, pertening to ye custom house of Hull'.
30.	Scarborough Castle		*Ms. Coloured plan.*
31.*	Durham	1576	*Accompanied by Justices.*
32.	Durham		*Ms. Plan of county. Pen and ink. A few additions by Burghley. See footnote 1.*
33.*	Northumberland		*Accompanied by Justices. In three margins of the map are, in Burghley's hand, the* 'Names of ye principall lordsh(ips) in ye Middle march (etc) and the Lordes names'. *Added are the names, etc., of the principal gentlemen with their places of residence in the* 'Est wardenrie of England', *the* 'middle marches of England', *the* 'Scottish Est marches', *and the* 'Scottish west marches'.
34.*	Westmorland, Cumberland	1576	*Accompanied by Justices and* 'Surnames of Gentlemen of the Middle Marches'.
35.	Shropshire		*Ms. plan. With principal towns only.*
36.	'A Platt of the opposete Borders of Scotland to ye west marches of England'	'Dec. 1599'	*Ms. plan. Title endorsed on map. Accompanied by a description of the map, ending:* 'for those on the English coast they ar referred to the tract latly sent to your L. of the description of them in particular'.
37.	A RETURN, ETC.	1592	*A return drawn up by Edward Aglionby of* (a) 'The division of the severall charge of the west borders of England and Scotland'; (b) *a list of places headed* 'The governaunce of Scotland most offensive to England lyeth in two words in Annerdale and Lyddisdale'; (c) 'Surnames of the English borderers and their dwellings', *and of the* 'borderers of Scotland'; (d) 'Names of the officers of the west wardenry of England'.

38.*	Lancashire	1577	*Accompanied by a list of Justices sworn on the 14th and 15th November, 1592, the list being signed 'Ed. Phytton' (? Sir Edward Fitton, the younger, father of Mary Fitton, who has been supposed to be the dark lady of Shakespeare's sonnets).*
39.	Lancashire		*Ms. plan. Pen and ink. Includes names of the gentry and their houses.*
40.*	Cheshire	1577	*Accompanied by: (a) Justices; (b) 'Postes . . . laid towards Ireland for hir Maties speedye and better service . . . in annis 1579, 1580 and 1581', at the various stages from London to Holyhead, with 'Rates of wages per diem'.*
41.*	Shropshire	1577	*Accompanied by Justices.*
42.	'Foresta de Morfe'		*Ms. Pen and ink drawing of the district east of Bridgnorth, Shropshire. Includes Bridgnorth and neighbouring villages.*
43.	Shrewsbury		*Ms. Coloured plan.*
44.	South-western Shropshire		*Ms. Coloured plan of the Clun district.*
45.	North-western Shropshire		*Ms. Coloured plan of the Oswestry district.*
46.*	Herefordshire	1577	*Accompanied by Justices and a short pedigree of Vaughan.*
47.*	Gloucestershire	1577	*Accompanied by Justices.*
48.	Wales		*Entitled: 'Cambriae Typus auctore Hvmfredo Lhvydo Denbigiense Cambrobritaño'. From Ortelius' 'Theatrum Orbis Terrarum', 1570.*
49.*	Monmouthshire	1577	*Accompanied by: (a) Justices; (b) names of landowners in Burghley's hand.*
50.*	Glamorganshire	1578	*Accompanied by Justices.*
51.*	Radnorshire, Breconshire, Cardiganshire, Carmarthenshire	1578	*Accompanied by: (a) Justices; (b) castles, parks and forests.*
52.*	Pembrokeshire	1578	*Accompanied by Justices.*
53.*	Montgomeryshire, Merionethshire	1578	*Accompanied by Justices, to some of whose names Burghley has added residences.*
54.*	Denbighshire, Flintshire	1577	*Accompanied by Justices.*
55.*	Anglesey, Caernarvonshire	1578	*Accompanied by Justices, etc.*
56.	Jersey		*Ms. Coloured plan. Endorsed: 'Jersey. The Haven'.*
57.	Scotland	1586	*Entitled: 'La vraye et entiere description du tresancien royaume pays et isles d'escosse'. An enlargement of the map which accompanies the work 'De origine moribus . . . Scotorum' by John Leslie, Bishop of Ross, published at Rome in 1578 (see footnote 2). At the foot of the map are a dedication to James VI in French, dated Rouen, 26th August, 1586, and a brief description of Scotland, also in French. 'Imprimé à Rouen 1586.'*

58.	Coasts of Norway (north of Trondheim), Lappland, and north-western Russia	1557 or later	*Ms. Map signed by William Borough, navigator and Comptroller of the Queen's Navy. Added is a table of distances from the Thames to St Nicholas, and from thence 'The Travaile vpp the Ryver of Dvyna' to Vologda and on to Moscow; the map only shows the first two stages from St Nicholas, Moscow being the twentieth.*

Footnotes
1. Described by Manley, Gordon. 'The earliest extant map of the county of Durham.' *Transactions of the Architectural and Archaeological Society of Durham and Northumberland*. Volume 7. 1936.
2. Described by Skelton, Raleigh A. 'Bishop Leslie's map of Scotland, 1578.' *Imago Mundi*. Volume 9. 1950.

Appendix 7

A LEASE OF LANDS AT GRIGSTON MANOR, SUFFOLK. 11th MARCH, 1574
(Public Record Office. Patent Roll. 16 Eliz., pt.14 (No.1121), m.34)

Abstract from Latin. Words in inverted commas are as given in the original. Those in brackets are added, being either the modern name of the places, or an unusual Latin phrase.

The Queen etc Know all that we, in consideration that Christofer Saxton for certain good causes grand charges and expenses lately had and sustained in the survey of divers parts of England, have sold granted and to farm demised to Christofer Saxton all those messuage lands etc as well of free as of customary tenants which late were assigned to lady Anne of Cleve for the term of her life in Stafford Benall Fernaham Great Glenham, Little Glenham Swefflinge, and Rendam in co. Suffolk (Stratford, Swefling, and Rendham) and which in the whole now extend to the annual rent of 9*li* 18*s* 3*d* and which are known by the name of 'Grigston lande' otherwise 'Grigston manor' by Stafforde and all that half acre of land called 'Rotton acre' of new enclosure (de nouo appruam^t) in Stratford of the yearly rent of 12*d*. and all manner of court leets and views of francpledge and the perquisits and profits of the same, also all houses, gardens, pastures, ponds etc to the said premises belonging, excepting, to us and our heirs, large trees, wood, underwood, wards, marriages, minerals and quarries, goods and chattels of felons and fugitives, advowsons of churches etc. To hold to Christopher Saxton his executors and assigns from Michaelmas last past for the term of 21 years next following, rendering yearly to us our heirs and successors 10*li* 5*s* 11*d* and he, the said Christopher, to do all necessary repairs. It shall be lawful for the said Christopher to have sufficient housebote, hedgebote, fyrebote, ploughbote and cartebote there during the said term, and he shall have timber from the woods on the premises for the repair of the mills, houses, and buildings during the said term. Etc.
Gorhamburye 11 March, 16 Elizabeth (1573–4).

Appendix 8

AN ORDER OF ASSISTANCE FOR SAXTON'S SURVEY OF WALES
10th JULY, 1576
(Public Record Office. Privy Council Register 2:11)

At St. James the xth of Julie 1576.

An open Lr̃e to all Justice of peace mayo^{rs} & oth^{rs} etc w^t in the severall Shieres of Wales That where (whereas) the bearer hereof Xp̄ofer Saxton is appointed by her Matie under her signe and signet to set forth and describe in coates(cartes) pticulerie all the shieres in Wales That the said Justices shalbe aiding and assisting unto him to see him conducted unto any towre Castle highe place or hill to view that countrey and that he may be accompanied wth ij or iij honest men such as do best know the cuntrey for the better accomplishement of that service and that at his depture from any towne or place that he hath taken the view of the said towne do set forth a horseman that can speke both Welshe and englishe to safe conduct him to the next market Towne, etc.

Appendix 9

A LICENCE FOR THE EXCLUSIVE PUBLICATION OF SAXTON'S MAPS
DURING A PERIOD OF TEN YEARS. 20th JULY, 1577
(Public Record Office. Patent Roll 19 Elizabeth, pt.9 (No.1159), m.21)

D' cōn p Xpofero Saxton

Elizabeth by the grace of god &c To all manner printers Booksellers and other oure officers Ministers and Subjectes gretinge Whereas Christofer Saxton servaunte to oure trustie and welbeloved Thomas Sekeford

Esquier Master of Requestes unto us hathe already (at the greate coste expenses and charges of his said master) traveyled throughe the greateste parte of this oure Realme of Englande and hathe to the greate pleasure and cōmoditie of us and our lovinge subjectes uppon the pfecte viewe of a greate number of the seu'all Counties and Sheires of oure said Realme drawen oute and sett fourthe diu'se trewe and pleasaunte mappes chartes or plattes of the same counties Together withe the Cities Townes Villages and Ryvers therein conteyned vearie diligentlye and exactly donne and extendithe yf god graunte hym lief further to travell therein throughe out all the residue of oure said Realme And so from tyme to tyme to cause the same plattes and discriptions to be well and fayre Ingraven in plates of copper and to be after Impressed and stamped out of the same aswell to the cōmoditie of oure subjectes as to all other that shall have pleasure to see and puse the same We lett youe witte that for the better incouraginge of the saide Christofer to pcede in this his so pfitable and beneficiall an enterprise to all manner of psons Of oure especiall grace certen knowledge and mere mocyon We have geven and graunted and by thies p'sentes doe giue and graunte Priviledge and licence unto the saide Christofer Saxton and to the assigne and assignes of hym that he the said Christofer Saxton and the assigne and assignes of hym onely And none other for and duringe the space of tenne yeres nexte ensewinge the date of this oure licence shall and maye by hym selfe his assigne and assignes factors and deputies Imprinte and sett fourthe or cause to be Imprinted and sett fourthe any and as manye suche mappes charttes and plattes of this oure Realme of Englande and Wales or of any Countye or other parte thereof by hym all ready or here after to be sett fourthe as to hym and to his saide deputie and deputies shall seme mete and conveniente And shall and maye sell or utter or cause to be solde or uttered any suche Imprinted Mappes Charttes or plattes as aforesaid And further we doe by theis p'sentes streightlye forbyd prohibite and cōmaunde all and singuler other parson and parsons aswell Printers and Booksellers as all and eu'y others whatsoever beinge either oure subjectes or straungers (other then the saide Christofer Saxton and the assigne and assignes of hym) or suche other as by the saide Christofer his executors or assignes shalbe appoynted That they nor any of them duringe the said terme of tenne yeres in any manner of wyse shall Imprinte or cause to be Imprinted drawen paynted or set fourthe any manner Mappe Charte or platte as aforesaide but onelye the saide Christofer Saxton or the assigne or assignes s'u^{a}nte or s'u^{a}ntes deputies or factors of hym the saide Christofer Saxton nor shall bringe in or cause to be broughte from the partes beyonde the seas into or withe in any oure Realmes or domynyons nor in the same shall sell utter or putto sale or cause to be solde uttered or putto sale or otherwyse disposed any of the saide Mappes Chartes or plattes of any oure Realmes or domynyons or any partes or parcelles of the same made or Imprinted in any forreyne countrye uppon peyne of oure hieghe indignacon and displeasure And that eu'y offender doinge contrarie to theffecte and true meaninge of thies p'sentes shall for eu'y suche offence forfeicte to thuse of us oure heires and successors the sōme of tenne poundes of laufull monye of Englande And shall also moreover forfeicte to the saide Christofer Saxton and to thassigne and assignes of hym all and eu'y suche Mappes Chartes and plattes as shalbe Imprinted solde or uttered contrarie to the trewe entente and meaninge of thies p'sentes Willinge therefore and cōmaundinge aswell the Master and Warders (sic) of the Misterye of Stacyoners in oure Cytye of London as also all other oure officers Ministers and subjectes as they will avoyde oure displeasure and indignacon that they and eu'y of them at all tymes when nede shall require duringe the saide terme doe ayde and assiste the saide Christofer Saxton and the assigne and assignes of hym and eu'y of them in the due exercysinge and execuōn of thies oure p'sente licence and priviledge withe effecte accordinge to the true meaninge of the same Althoughe expresse mencion &c. In witnes whereof &c. Witnes oure selffe at Goramburie the two and twentie daye of Julye (19 Elizabeth, 1577).

p p̄iam Reginam &c.

Appendix 10

EXTANT COPIES OF CHRISTOPHER SAXTON'S ATLAS OF ENGLAND AND WALES (1579) IN THE BRITISH ISLES AND UNITED STATES OF AMERICA

Certain copies of Saxton's Atlas have been lost or misplaced over time; these are recorded at the end of the appendix. Privately held copies are, at the request of their owners, omitted for the most part; those listed represent but a few of several known to exist, and the details documented which pertain thereto have been either supplied willingly by the owners or obtained from previously published work. A few copies held in libraries have been omitted by request. The collection of proof maps which belonged to Lord Burghley (British Library. Royal Manuscripts, 18.D.III) is excluded from the list. Plates are very occasionally missing from some copies, but such omissions are not documented.

Location	Reference	Frontispiece (State I/II)	Index (Setting A, B, C, D)	Engraving of Coats-of-Arms/ Table of Cities	Map of 'Anglia...' (State I/II)	Map of Northants. (State I/II)	Map of Norfolk (State I/II)	Documented References	Notes
BRITISH ISLES *Special Libraries*									
1. Aberystwyth. National Library of Wales.	cupboard in annexe to Print Room	wanting	wanting	wanting	I	I	I		
2. Dublin. Chester Beatty Library.	12–4	II	D	present	II	II	II		
3. London. British Library.	Dept. of Prints & Drawings. 172.d.4	II	D	present	II	II	II	Chubb VI. Hodson 1.	
4.	Map Room. c.7c.1	II	D	present	II	II	II	Chubb I. Hind p.88. Hodson 1. Skelton 1 + p. 209.	Bound up with this copy, and printed on the same paper, are Ryther's plates of the Spanish Armada (*Expeditionis Hispanorvm In Angliam vera descriptio*), engraved to accompany P. Ubaldini's *Discourse* (1590). This atlas probably belonged to James I. Chubb wrongly gives the Index as wanting.
5.	Map Room c.7c.2	wanting	B	wanting	I	I	I	Chubb I. Hind p.89. Hodson 1. Skelton 1.	One of earliest known copies of the atlas. Frontispiece wanting, but its position occupied by a blank leaf of the original paper. Map of 'Anglia...' occurs twice in both instances in State I.
6.	North Library. G.3604	II	D	present	II	II	II	Chubb V. Hodson 1.	Appended at the end of the volume is the Quartermaster's Map by Wenceslaus Hollar (1644) and ten other maps.

7.	North Library. 118.e.1	II	A	wanting	I	I	I	Chubb III. Hind p.89. Hodson 1. Skelton 1 + p. 209.	
8. London. The Library, Inner Temple.	no index nor Catalogue number	I	B	wanting	II	I	I		This copy also contains illustrations of twenty-nine English cities.
9. London. The London Library.	Safe 4, folio	II	D	present	II	II	II		
10. London. National Maritime Museum.	SAX 001	I	D	present	II	I	I	Hind p.74.	
11.	SAX 002	facsim of II	D	present	II	II	II		
12. London. Royal Geographical Society.	264.H.12	II	D	present	II	II	II	Hodson 1.	
13. London. Society of Antiquaries of London.		II	D	present	II	II	II		
14. Manchester. Chetham's Library.	3501	I	B	wanting	II	I	I		Originally the copy belonging to C. Hatton.
15. York. York Minster Library.	(S)III.A.8.	II	D	present	II	II	II		

Municipal Libraries

16. Birmingham. Birmingham Reference Library.	86973	II	D	present	II	II	II		
17.	307454	II	D	present	II	II	II		Following the preliminary pages there is '*a mapp of England printed anno 1652*', printed and sold by P. Stent.
18. Leeds. Central Library.	SRF 912. 42 SA 98	II	D	present	II	II	I		

University of Cambridge								
19. University Library.	Atlas. 4.57.6	II	D	present II		II	II	Hodson 1.
20. Fitzwilliam Museum.	AA54	II	D	present II		II	II	
21. Peterhouse.	S 55	I	A	wanting I		I	I	This copy belonged to Andrew Perne, Master of Peterhouse, 1553/4–89. Its cartographic features indicate an early issue.
University of Oxford								
22. Bodleian Library.	Douce Prints. b.27	II	D	present II		II	II	Seven additional maps are appended to the atlas, including Humphrey Lhuyd's '*Angliae Regni Florentissimi Nova Descriptio, Avctore Hvmfredo Lhvyd Denbygiense 1573 Cum Priuilegio*' and '*Cambriae Typus Auctore Hvmfredo Lhvydo Denbigiense Cambrobritaño*'.
23.	Fol. B.S. 45	II	D	present II		II	II	
24.	Gough Maps. 96	wanting	wanting	wanting II		II	II	Originally possessed Frontispiece (State II), Index (Setting D), and plate of Coats-of-Arms/Table of Cities.
25. Balliol College Library.	925 b 8	wanting	D	wanting I		I	II	
26. Christ Church Library.	Evelyn Collection d.153	wanting	wanting	wanting wanting		I	I	See note at end of table.
27. New College Library.	Ω 45.01	I	B	wanting I		I	I	The atlas bears the stamp of Sir John Savile (1545–1607), one of the Barons of the Exchequer. The pages are in an extremely early state. A number of manuscript additions occur.

28. Queen's College Library.	Sel.b.223	wanting	D	present		II	II	II	
Other Universities									
29. Dublin. Trinity College Library, University of Dublin.	M.aa.3	wanting	D	present	wanting	II	II		
30. Glasgow. University Library, University of Glasgow.	Hunterian Collection Di.l.12	wanting	D	wanting	I	II	II		
31. Leeds. Brotherton Library, University of Leeds.	Whitaker Collection 1	II	D	present	II	II	II		
32. Liverpool. Harold Cohen Library, University of Liverpool.	H49.45	II	D	wanting	I	II	II		Several ms. additions are interpolated into the atlas.
33. Manchester. John Rylands University Library.	R 1771	II	D	present	II	II	II		
34. Wales. University College of North Wales Library, Bangor.	912.42	II	wanting	present	II	II	I		
Private Collections									
35. Sir Hugo Boothby, Fonmon Castle, South Glamorgan.		wanting	wanting	wanting	I	II	I		
36. Lord Crawford, Balcarres, Colinsburgh, Fife.		I	A	wanting	I	I	I	Hind p.74. Skelton	Early version of Frontispiece which probably, though not certainly, belongs to the atlas. Variant I of early setting of Index, containing five misprints. Almost certainly one of the earliest issues of Saxton's Atlas.

UNITED STATES OF AMERICA									
37. Chicago. Newberry Library.	Case+ G1045. 78	II	D	present	I	II	II	Skelton 1.	A table of distances in Yorkshire, taken from John Norden's '*A Direction for the English Traviller, 1635*', is inserted into the atlas, but is probably a late addition.
38. New York. Pierpont Morgan Library.	33483	II	D	present	II	II	II		Inserted into the atlas is a map entitled: '*A general Plott and description of the Fennes and surrounded grounds in the sixe Counties of Norfolke, Suffolke, Cambridge . . . Henrici Hondii. 1632*'.
39. San Marino, California. Henry E. Huntington Library & Art Gallery.	82871	II	D	present	I	II	I	Skelton 1.	
40.	110105	II	D	present	II	II	I		
41. Washington D.C. Folger Shakespeare Library.	STC 21805.5	II	D	present	II	II	II		
42. Washington D.C. Library of Congress.	G1807.S3 1579. Map Cage. Copy 1.	II	D	present	II	II	II	Phillips 2913	
43.	G1807.S3 1579. Map Cage. Copy 4.	II	A&D	wanting	I	I	I	Phillips 8109	
44. Library of Congress. Lessing J. Rosenwald Collection. Jenkintown, Pennsylvania.	G1807.S3 1579. Map Cage. Copy 2.	II	C	present	I	II	II	Hind p.89. Phillips 8108. Skelton 1.	Once in the possession of Francis Walsingham, a secretary of state during reign of Elizabeth I. Still in its original calf binding.

45. Library of Congress. Lessing J. Rosenwald Collection. Jenkintown, Pennsylvania.	G1807.S3 1579. Map Cage. Copy 3.	I	wanting	wanting	I	II	II	Hind p.74. Phillips 8107. Hodson 1. Skelton 1.	Only known issue of Saxton's Atlas to have been printed on vellum. Must therefore have been printed as an entity, and clearly represents a very early issue of the atlas. Despite rarity, Frontispiece also belongs to atlas as a whole.
46. Urbana-Champaign. University of Illinois Library.	uncatalogued	I	D	present (facsim.)	II	II	II		
47.				Hind (1952) lists a copy in the Houghton Library, University of Harvard. This copy once belonged to John Camp Williams. It possesses the Frontispiece, Index, and maps, all in their final state, as well as the plate of Coats-of-Arms/Table of Cities. The copy, in the precincts of the Houghton Library, is actually privately owned, but the owner wishes to remain anonymous.					

Note
Copy 26. Christ Church Library. Evelyn Collection d.153. This atlas was sold at Christie's, March 1978, Lot 1303 for £18,000. It was bought by Deemond Burgers, a dealer, and subsequently dispersed.

Lost Copies
A. Signet Library, Edinburgh.   Sold 1960.
B. Gray's Inn Library, London.   Lost, together with bibliographical records, owing to war damage.
C. Lambeth Palace Library, London.   Lost, presumed stolen, mid-1970s.
D. Sion College, London.   Catalogue entry for Saxton's 'CHARTAE GEOGRAPHICAE OMNIUM COMITATUM ANGLIAE ET WALLIAE COLORIBUS DISTINCTAE' (1579). Catalogue No.ARC B 40. 1/SA 9. Disappeared.
E. R. A. Nicholson.   Offered to British Library, but declined on grounds of similarity to North Library, 118.e.1. copy. Frontispiece (State II); Index (Setting A, with five misprints); Norfolk and Northamptonshire (both in State I); wanting 'Anglia ...' and plate of Coats-of-Arms/Table of Cities. Subsequently dispersed.

References
Chubb, Thomas. *The Printed Maps in the Atlases of Great Britain and Ireland: A Bibliography, 1579–1870.* (1927).
Hind, Arthur M. *Engraving in England in the Sixteenth and Seventeenth Centuries. A Descriptive Catalogue with Introductions. Part I. The Tudor Period.* (1952).
Hodson, Donald. *The Printed Maps of Hertfordshire, 1577–1900.* (1974).
Phillips, P. L. *A List of Geographical Atlases in the Library of Congress.* (1909-20).
Skelton, Raleigh A. *County Atlases of the British Isles 1579–1850. A Bibliography.* (1970).

Appendix II

TITLES AND SEQUENCE OF MAPS APPENDED AT THE END OF THE COPY OF SAXTON'S ATLAS IN THE BRITISH LIBRARY
(North Library; G.3604)

1. The / Kingdome / Of / Scotland.
2. The / Kingdome / Of / Irland / Devided into severall Provinces, and thē / againe devided into Counties. / Newly described. / ... Performed by Iohn Speede and are to be sold by / Roger Rea the Elder and younger at / the Golden Crosse in Cornhill against / the Exchange 1662 / ... Jodocus Hondius caelavit.
3. The / Province Of / Connaugh / with the Citie of Galwaye / Described. / ... Performed by Iohn Speede and are to be solde by Roger Rea the / Elder and younger at the Golden Crosse in Cornhill / against the Exchange / ... Cum / Privilegio / Anno / Domini / 1610.

4. The / Countie Of / Leinster / with / The Citie Dublin / Described / ... Performed by John Speed, and are to be sold / by Roger Rea the Elder and younger / at the Golden Crosse in Cornhill against / the Exchange / ... Anno Domini 1610. / Iodocus Hondius caelavit.
5. The / Province / Of / Mounster / ... Performed by Iohn Speede and are to besolde by Roger / Rea the Elder and younger at the Golden / Crosse in Cornhill against the Exchange / ... Jodocus Hondius Caelavit.
6. The Province / Ulster / described. / ... Performed by John Speede and are to be solde / by Roger Rea the Elder and younger at the / Golden Crosse in Cornhill against y^e Echang.
7. A Mappe Of Kent Sovthsex Svrrey Mid = / dlesex Barke and Southampton Shire & the Ile / of Wight part of Essex & Wiltshire etc. / ... Printed Coloured and / Sold by Iohn Garrett at the South Entrance / of the Royall Exchange in London.
8. The Mappe Of Norfolke, Svffol = / ke, Cambridgeshire, Bedford, Hartford, Buc = / kingham, Oxford, Northāpton, Warwick, Huntĩgtō, and Lecester Shires & Rutland, part of Lincolne, Nottingh⁻, / Darbye, Glocester, & Barck Shires, & of the County of Essex.
9. The Mappe of Shropshire / Cheshire Staffordshire Wostershire / Herefordshire, Brecknokshire, Caermar / thinshire Cardiganshire Radnorshire / Montgomeryshire Den / bighshire, Flyntshire, Carnarvan, the Ile / of Anglesey, & parte of Pembrokeshire.
10. The Bishop = ricke Of Dvrram / And Cvmberland, Westmoreland, / Yorke = shire, Lancast = shire, And / Parte Of Linconshire.
11. A Mappe Of Penbrokeshire / Glamorganshire, Monmouthshire, Glo = / ster shire, Somersetshire, Dorsetshire, / Devonshire, & Cornwall, part of Wilt: / shire, etc ... Carte de L'Angleterre et d'une Partie d'Ecosse Gravée par le / Celebre Hollar. / Cette Carte a été Réduite d'après les Provinces de M^R. Saxton Par ordre / d'Oliver Cromwells Pour l'usage de ses Armées, ce qui fait qu'elle est / Réputée pour être la Meilleure de l'Isle : Toutes celles qui ont été publiées depuis n'en / sont que des Copies. elle est connuë Sous le nom de Quarter Master's Map / ou Carte des Quartiers Maîtres; les Planches ont été dans l'oubly / pendant plusieurs années par le peu de connoissance de Celuy qui le Possédoit. / elles ont été acquises par le S^R. J. Rocque Chorograph e de S.A.R. / Monseigneur le Prince de Galles, dans le Strand A Londres 1752 / le Prix est d'une demy-Guinée ou 12th.

Appendix 12

THE PROJECTED EDITION OF 1665
DETAIL OF CARTOGRAPHIC MODIFICATIONS
TO THE CONSTITUENT COUNTY MAPS

Each map has a preceding state in the Web edition of 1645, and a succeeding state in the Lea edition of c.1689.

1. Cornwall — Saxton's Latin title and its cartouche, which survived in the Web edition, are replaced by a plan of 'Lavnceston or Ancient Dunhevet', and a view of 'The Hurlers', together with 'The Chesewring' and 'The other Halfstone', all after Speed. The Royal Arms of Charles I, and the royal cypher 'C.R.', as in Web's edition, remain above the plan of Launceston. Seckford's arms are replaced by eight heraldic shields with 'The Armes of Persons Dynĩfyed wth ye title of Cornwall', and reference letters are added outside the map borders.

2. Derbyshire — Seckford's arms are replaced by a plan of 'Darbie', after Speed, which obliterates the lower part of Saxton's original title cartouche. Other additions are 'Bvxton' (a view of St Ann's Well, after Speed), 'The Tiding Well', and 'ye Deuills arse in ye Peake'. Two heraldic shields are added between the English title and the county boundary, i.e. 'William Ferres, Ed. E. of Lanca:' and 'Thomas Standley', and two arms of the Earl of Chesterfield are added on the right-hand side of the map. The cypher 'C.R.' is added above the Royal Arms of Charles I, which survive from Web's edition. Hundreds have been inserted, reference letters added outside the map borders, and some place-names and hills in adjoining counties also appear.

3. Dorset — The Royal Arms of Charles I (from the Web edition) and Saxton's original Latin title and its cartouche are replaced by a plan of Dorchester, after Speed. The Royal Arms of Charles I are engraved to the right of this plan. Seckford's arms are removed, and replaced by the new English title. Four heraldic shields appear in the bottom right of the map, and hundreds have been added. Additional place-names appear in adjoining counties, and reference letters are added outside the map borders.

4. **Durham** Saxton's title-cartouche, containing Web's new English title, is replaced by a plan of Durham, after Speed. This plan obliterates the letters 'NORTH', which have therefore been re-engraved above the remainder of this name. Seckford's arms are replaced by a new English title and a small title-cartouche. The original date (1576) remains below the scale in the bottom right-hand corner. The Royal Arms of Charles I, and the royal cypher 'C.R.', survive from the Web edition. Reference letters appear outside the map borders, and place-names and hills in adjoining counties have been added.

5. **Glamorganshire** No significant changes have occurred to the general format of the map, but the royal cypher 'C.R.', and hundreds, have been added.

6. **Gloucestershire** Saxton's original title-cartouche is replaced by a plan of 'Brestoll', after Speed. The original scale is replaced by a plan of 'Glocester', again after Speed, and by eleven heraldic shields; a copy of Saxton's scale is therefore engraved in the top right-hand corner. The Royal Arms of Charles I and cypher 'C.R.' survive from the Web edition, and surmount the plan of Bristol. The new title, which is uncertain, replaces Seckford's arms. Hundreds have been added, and extra place-names appear in adjoining counties.

7. **Hampshire** The Royal Arms of Elizabeth I, which survived from Saxton's original plate on Web's map, together with Saxton's Latin title and its cartouche, are replaced by a plan of Winchester, after Speed. Five heraldic shields are added beneath the scale, and four more (one of which is blank), together with the Royal Arms of Charles I, replace Seckford's arms in the lower right. Saxton's open dividers are removed, but the line-scale itself remains unaltered. The title, (which is uncertain), has been removed to occupy the cartouche originally containing the index on the right. Hundreds have been added, reference letters appear outside the map borders, and extra place-names have been inserted in adjoining counties.

8. **Herefordshire** A new English title replaces Saxton's Latin title, and the cypher 'C.R.' is added above the Royal Arms of Charles I, which survives from the Web edition. Six heraldic shields are added to the map, and the hundreds are inserted. Reference letters are added outside the map borders, and additional place-names occur.

9. **Lancashire** Seckford's arms are replaced by a plan of 'Lancaster towne & Castle', after Speed. Two heraldic shields occur below this plan (of the Earl of Manchester, and a blank shield intended for the arms of the Earl of Warrington). Heraldic shields containing 'The Arms of ye families y^t have borne ye titles of Lancaster' replace a large ship, occurring above the Royal Arms of Charles I (which survives from the Web edition). The royal cypher 'C.R.' is added. Hundreds have been added, and reference letters appear outside the map borders. Place-names have been added in adjoining counties.

10. **Monmouthshire** A plan of 'Monmovth', with its arms, (after Speed), is inserted at the lower right of the map. Seckford's arms are replaced by the Prince of Wales' plume of feathers, the arms of 'Robert Cary Ea: of Monmouth', a cherub, and a blank shield for the 'Du: of Monmouth'. The royal cypher 'C.R.' is added above the Royal Arms of Charles I, which survive from the Web edition, and which surmount the new English title. Hundreds have been added, and reference letters appear outside the map borders. Extra place-names have been added in adjoining counties.

11. **Norfolk** Saxton's Latin title and its cartouche are replaced by a new English title in a new cartouche. Saxton's index and its cartouche, in the top right-hand corner, are replaced by a plan of Norwich, after Speed. The Royal Arms of Elizabeth I are replaced by seven heraldic shields, while the Royal Arms of Charles I replace Seckford's arms. The original hundred-boundaries have been reworked, and the names of the hundreds added. New scale numerals have been engraved, and reference letters appear outside the map borders.

12. Northamptonshire	Saxton's original Latin title and its cartouche, and the Royal Arms of Charles I, are replaced by a plan of Northampton, after Speed, surrounded by heraldic shields of 'Families dignifyed with ye title of Northampton'. The new English title, in a new cartouche, has been inserted at the bottom right of the map. Below the plan of Northampton occurs a small plan of 'Peterburgh', again after Speed, and in this respect akin to the other additional plans of 'Oukham', 'Hvntington', and 'Bedforde', the last of which replaces Seckford's arms. The Royal Arms of Charles I have been re-engraved in the bottom right-hand corner, and possess the initials 'C.R.'. Hundreds have been added, and extra place-names appear in adjoining counties.
13. Shropshire	A plan of Shrewsbury, after Speed, surmounted by two shields supported by cherubs, replaces Seckford's arms. The new English title is surmounted by the Royal Arms of Charles I, which survive from the Web edition. Hundreds have been added, and reference letters appear outside the map borders. Extra place-names and rivers have been added in the adjoining counties.
14. Somerset	Saxton's Latin title and its cartouche, together with the Royal Arms of Charles I (after Web), are replaced by a plan of 'Bathe', after Speed, and by two heraldic shields (the Earls of Bath and the Earls of Bridgwater). The new English title is engraved at the top-centre of the map, and lies above the newly engraved Royal Arms of Charles II. Seckford's arms are replaced by five heraldic shields bearing the arms of 'Families bearing ye title of Dukes and Earles of Somersett'. Hundreds have been added, and reference letters appear outside the map borders. Extra place-names have been added in Devonshire.
15. Staffordshire	A plan of Stafford, after Speed, replaces the Royal Arms of Elizabeth I, and a plan of Lichfield, again after Speed, replaces Seckford's arms. The Royal Arms of Charles I, and the cypher 'C.R.', are inserted on the left of the map. Below Saxton's surviving scale there is a blank shield. Hundreds have been added, and reference letters appear outside the map borders. Extra place-names have been inserted in adjoining counties.
16. Suffolk	Saxton's Latin title and its cartouche, and the Royal Arms of Charles I, are replaced by a plan of Ipswich, after Speed. The new English title has replaced Seckford's arms. The Royal Arms of Charles I are re-engraved in the bottom left-hand corner, still without the royal cypher 'C.R.', and heraldic shields occur to the left and right of the plan of Ipswich. Saxton's hundred boundaries have been re-engraved, while reference letters have been added outside the map borders.
17. Wiltshire	The Royal Arms of Charles I, Saxton's Latin title and its cartouche, and Seckford's arms, are eliminated. They are replaced by a plan of 'Salisbvry' and a view of 'Stone heng', both after Speed, and by eleven heraldic shields bearing the arms of 'The Earles of Wiltshire' and 'The Earles of Salisbury'. The new English title is interpolated between the plan of Salisbury and the view of Stonehenge. Inserted on the right-hand side are the arms of the Earls of Carendon and Marlborough, and the Royal Arms of Charles I (with the cypher 'C.R.'). Hundreds have been added, reference letters appear outside the map borders, and a few extra place-names occur.
18. Worcestershire	Saxton's Latin title and its cartouche, which survived in the Web edition, is replaced by a plan of 'Worcester Citty', after Speed, and heraldic shields are inserted beneath this. The new English title replaces Seckford's arms. Hundreds have been added, and reference letters appear outside the map borders. Extra place-names in adjoining counties have been added.
19. Yorkshire	Few changes have occurred to the general format of the map, but it remains in two sheets (as in the Web edition), and it still lacks a title. The names of the wapentakes, however, have perhaps been added.

Appendix 13

LEA'S EDITIONS OF *c*.1689 AND *c*.1693
CARTOGRAPHIC MODIFICATIONS TO THE MAPS

Maps are listed alphabetically. The preceding state of each map is indicated in parenthesis, details being given in Chapter 5a for the Web edition of 1645 and in Appendix 12 for the edition of 1665. The maps of Devonshire and Northumberland which do not derive from, but which are copies of, Saxton's plates, are included.

1. **Anglesey**
   (preceding state, Web; 1645)

   (a) *c*.1689. A new English title and title-cartouche replace Seckford's arms. The Royal Arms of Elizabeth I and Saxton's Latin title, both of which survived in Web's edition, are erased, and the spaces left blank. The hundreds are added, but only in Anglesey. A crown, mitre, and crosses are added to those towns to which they are applicable.

   (b) *c*.1693. A plan of Beaumaris occupies the position originally taken by the Royal Arms erased in *c*.1689. The vacant spaces once occupied by Saxton's title and its cartouche are taken by plans of Bangor and Caernarvon. The hundreds have been added in Caernarvonshire. Roads are shown throughout the map, and a few heraldic shields are added.

2. **'Anglia...'**
   (preceding state, Web; 1645)

   (a) *c*.1689. The date 1642, which occurred in Web's English title, is erased, and Lea's imprint is added at the foot of the cartouche, i.e. 'P. Lea excudit'. Saxton's Latin 'Index Omnivm Comitatvvm' now reads 'A Catalogue containing all ye Shires', and contains ten columns to the right of the county names, giving the numbers of cities, bishoprics, markets, castles, parishes, rivers, bridges, etc., in each county. The elaborate frame with which Saxton enclosed the 'Index' is also replaced by a plain frame enclosing the 'Catalogue'. The Royal cypher 'C.R.' (after Web) is altered to 'W.R.'. The small cartouche containing a general note on the map (on the left of Web's plate), the panel containing Augustine Ryther's name, and Seckford's arms, are all erased, and are never subsequently replaced.

   (b) *c*.1693. No further changes of format occur, although 'By Chr. Saxton' is added below the title.

3. **Cheshire**
   (preceding state, Web; 1645)

   (a) *c*.1689. The Royal Arms of Charles I and the English title (after Web, where it was contained in Saxton's original cartouche) are replaced by a plan of Chester, and by a scroll containing references to this plan (the latter has necessitated the re-engraving of 'LANCASTRIAE PARS'). The new English title, occupying a piece of drapery nailed at two corners, replaces Seckford's arms. Hundreds have been added, and crosses to market towns. A few extra place-names have been added within the county boundary, e.g. 'Harbridge' and 'Lache', and eight heraldic shields also appear (two of which are blank), replacing the name of Francis Scatter and the date 1642.

   (b) *c*.1693. No further changes of format occur, but roads are added, and Chester possesses a crown and mitre.

4. **Cornwall**
   (preceding state, 1665)

   No significant changes of format occur, most having been effected in 1665. The only modifications are:
   (a) *c*.1689. The addition of crosses for market-towns.
   (b) *c*.1693. The re-engraving of the title, the addition of roads, and the application of crowns to borough towns.

5. **Denbighshire**
   (preceding state, Web; 1645)

   (a) *c*.1689. Seckford's arms have been erased, and the space left blank. A new English title occupies the position where, on Web's plate, Saxton's Latin title still appeared, but this remains in Saxton's original cartouche; this is still surmounted, as in Web's edition, by the Royal Arms of Charles I. A crown, mitre, and crosses have been added to the appropriate towns, and 'Offa's Ditch' is depicted.

   (b) *c*.1693. The erasure in *c*.1689 of Seckford's arms has permitted the insertion of a plan of Denbigh, while Saxton's scale, which still appeared in *c*.1689, is now erased and replaced by plans of St. Asaph and Flint. A new scale is accordingly engraved to the left of the title and Royal Arms, both of which survive from the edition of *c*.1689. Hundreds and roads have been added.

6. **Derbyshire** (preceding state, 1665)	No significant changes of format occur, most having been effected by 1665. By c.1689 a crown had been added to Derby and crosses to market-towns, and Lea's imprint had replaced the date in the title-cartouche. By c.1693 the title had been re-engraved and roads added.
7. **Devonshire** (preceding state, Web; 1645)	Saxton's plate has disappeared, and a copy (though lacking the marginal cardinal points) by Francis Lamb takes its place, revealing important changes of format. By c.1689, the spaces occupied in Web's edition by Seckford's arms, Saxton's Latin title and its cartouche, and Saxton's scale, are taken by a plan of 'Excester'. A new English title appears in the top-right (engraved as if upon a tapestry hanging from the two upper corners), a new scale in the bottom-left (replacing the compass-rose), and ten heraldic shields above this scale. The Royal Arms do not appear, and Saxton's cardinal points in the border are omitted. No further important changes of format in the edition of c.1693 are apparent, although roads are added and crosses, crowns, and a mitre are appended to the appropriate towns.
8. **Dorset** (preceding state, 1665)	No significant changes of format occur, most having been effected by 1665, but crosses and crowns had been added by c.1689, and by c.1693 the title had been re-engraved and roads added.
9. **Durham** (preceding state, 1665)	No significant changes of format occur, most having been effected by 1665. By c.1689, however, a mitre had been added to Durham and crosses for market-towns, and by c.1693 the title had been re-engraved, main roads inserted, and the county divided into four wards.
10. **Essex** (preceding state, Web; 1645)	Saxton's plate, missing in the edition of c.1689, is replaced by a new map by Francis Lamb. By c.1693, however, it reappeared, with its format appreciably modified. The new English title had been engraved upon the upper part of the ornamental cartouche used by Saxton (which had survived in the Web edition), and beneath this, replacing the lower part of Saxton's original cartouche (containing Seckford's arms) is a plan of Colchester and eight heraldic shields (one of which, 'Arthur Capell. E.', is blank). The Royal cypher 'W.R.' had replaced 'E.R.' above the Royal Arms of Elizabeth I, although the latter remained un-altered. Reference letters had been added outside the map borders (possibly suggesting that this map merits inclusion in the projected edition of 1665?), crosses and crowns have been added to the appropriate towns, and roads appear.
11. **Glamorganshire** (preceding state, 1665)	(a) c.1689. Seckford's arms have been erased and the space left blank. Crosses have been added to market-towns. (b) c.1693. A plan of 'Cardyfe' has been inserted in the space previously occupied by Seckford's arms in the Web edition, and a plan of 'Landaffe' has been inserted in the bottom right-hand corner. A crown has been added to Cardiff and a mitre for Llandaff, and roads appear.
12. **Gloucestershire** (preceding state, 1665)	No significant changes of format occur, most having been effected by 1665. By c.1689, roads had been added, and crosses, crowns, and mitres to the appropriate towns. By c.1693, road destinations were given, e.g. 'from Montgomery' and 'to London', and compass-indicators added to the two plans of Bristol and Gloucester (though the Hodson copy represents an intermediate state, lacking these compass-indicators).
13. **Hampshire** (preceding state, 1665)	No significant changes of format occur, most having been effected by 1665. By c.1689, roads had been added, and crowns, crosses, and mitres to the appropriate towns. The only addition on the map of c.1693 were the arms of 'Cha. Fitz Roy D' to a formerly blank shield, in the bottom left-hand corner.
14. **Herefordshire** (preceding state, 1665)	No significant changes of format occur, most having been effected by 1665. This date survives in the edition of c.1693. By c.1689, however, crosses, crowns, and a mitre had been added to the appropriate towns, and by c.1693 roads had appeared.
15. **Hertfordshire** (preceding state, Web; 1645)	(a) c.1689. The Royal Arms of Elizabeth I, and Saxton's Latin title, all of which survived in the Web edition, are erased, and the space left blank. The new English title, in a new cartouche, replaces Seckford's arms at the

bottom-right. Crosses and crowns have been added to the appropriate towns.

(b) *c*.1693. The position previously occupied by the Royal Arms and Saxton's title has been taken by a plan of 'Hartforde' and three heraldic shields, the one on the extreme right of the row of three being blank. Roads have been added.

16. Kent
(preceding state, Web; 1645)

(a) *c*.1689. The new English title has been added at the top of the map, inserted between the Royal Arms and the old title-cartouche in the top right-hand corner, and is engraved as a hanging tapestry. The small cartouche at the bottom of the title-cartouche, originally containing the dedication, is blank. Crosses, crowns and mitres have been added to the appropriate towns, and the divisions of lathes, rapes, and hundreds appear.

(b) *c*.1693. The small cartouche originally containing the dedication, left blank in the edition of *c*.1689, has been filled in by Lea's imprint. The roads and five heraldic shields have been added.

17. Lancashire
(preceding state, 1665)

No significant changes of format occur, most having been effected by 1665. By *c*.1689, crosses had been added for market-towns, and crowns for Liverpool, Manchester, Wigan, Preston, Clitheroe, and Lancaster. A second state of the edition of the map of *c*.1689 also exists, showing the addition of roads. By *c*.1693, the crown incorrectly applied in *c*.1689 to Manchester had been erased, and the title re-engraved.

18. Lincolnshire
(preceding state, Web; 1645)

(a) *c*.1689. The Royal Arms of Charles I, the English title (in Saxton's original cartouche), and Seckford's arms, all of which had survived in the Web edition, have been replaced by a plan of 'Lincolne', with heraldic shields, and by a new title, again in English. The two large ships in the North Sea have been erased (together with the sea-monster in the Wash and all but one of the small boats), and the spaces left blank. Crosses have been added for market-towns, a crown and mitre for Lincoln, and a crown for Stamford.

(b) *c*.1693. No further important changes of format have occurred, but crowns have been added to Nottingham and Boston, and roads appear.

19. Monmouthshire
(preceding state, 1665)

No significant changes of format occur, most having been effected by 1665. By *c*.1689, crosses had been added to the market-towns, and by *c*.1693 the roads had appeared and the arms of James Scott (Du. of Monmouth) completed (the right-hand shield of the two).

20. Montgomeryshire
(preceding state, Web; 1645)

(a) *c*.1689. Saxton's Latin title and its cartouche, together with the Royal Arms of Elizabeth I, all of which had survived in Web's edition, have been erased, and the space left blank. The new English title replaces Seckford's arms. A ship and sea-monster in the bottom-left have been erased, and the space left blank. Hundreds appear, and crosses have been added to market-towns.

(b) *c*.1693. The space previously occupied by Saxton's title and the Royal Arms of Elizabeth I has been filled by a plan of Montgomery, while a plan of Harlech occupies the bottom-left corner, the erasures of *c*.1689 clearly having been made with this in view. The arms of the town of Montgomery appear above the new title added in *c*.1689, further hundreds have been added, and roads appear.

21. Norfolk
(preceding state, 1665)

No significant changes of format occur, most having been effected by 1665. By *c*.1689, the date 1665 had been erased (though it is still visible), a mitre had been added to Norwich, and crosses and crowns to the appropriate towns. By *c*.1693 the title had been re-engraved, so that the date '1665' has become invisible, the roads added, and the arms of 'Hen. How. E.' inserted.

22. Northamptonshire
(preceding state, 1665)

No significant changes of format occur, most having been effected by 1665. By *c*.1689, a mitre had been added to Peterborough, and crowns and crosses to the appropriate towns. Traces of an erased inscription appear below the new title-cartouche. By *c*.1693 the roads had been added.

23. Northumberland
(preceding state, Web; 1645)

Saxton's plate disappears after the Web edition, but is replaced by a copy by an anonymous cartographer. The most important changes in the format, as indicated by this copy, had occurred by *c*.1689. The plan of 'Barwick' (after Speed) appears in the same position as it did in Web's edition. A plan of 'NEWE-CASTLE' (after Speed), occupies the position of the Royal Arms of Charles I, and the English title and its cartouche, as they occurred in the Web edition. Roman altars occupy the previous positions of Seckford's arms and Saxton's scale, a new plain scale (without dividers) is engraved at the bottom, and nine heraldic shields are added. Crowns and crosses are appended to the appropriate towns, and roads appear. The title is engraved at the top-centre of the map. Saxton's cardinal points, printed in Latin, are given in English. The Bodleian copy (Douce Prints b.28) represents an unfinished plate, lacking the town-plans and the heraldic shields. By c.1693, no further important changes of format had occurred, but the wards had been added.

24. **Oxfordshire** (preceding state, Web; 1645)
    (a) c.1689. A plan of Oxford replaces Saxton's Latin title and its cartouche, and the new English title replaces Seckford's arms in the bottom left-hand corner. The Royal Arms of Elizabeth I, which survived in Web's edition, has been erased from the top left-hand corner.
    (b) c.1693. No further significant changes of format have occurred. The roads are added, and the arms of 'Fra. Noris' and 'Tho. Howard', Earls of Berkshire, have been added, both above the title-cartouche.

25. **Pembrokeshire** (preceding state, Web; 1645)
    (a) c.1689. The Royal Arms of Elizabeth I and Saxton's Latin title and its cartouche, which had survived in Web's edition, have been replaced by a plan of Pembroke and eleven heraldic shields. Seckford's arms have been erased, but the space left blank. Saxton's scale survives, and the new English title has been inserted in a small cartouche above it. Hundreds are shown, and crosses, crowns, and a mitre have been added.
    (b) c.1693. The previous erasure of Seckford's arms has allowed the addition of a plan of Haverfordwest.

26. **Radnorshire** (preceding state, Web; 1645)
    (a) c.1689. The Royal Arms of Elizabeth I and Saxton's Latin title and its cartouche, all of which survived in the Web edition, have been erased and the space left blank. The new English title replaces Seckford's arms, and there is a small erasure in the bottom right-hand corner, which deletes a part of the 'Rumney flu'. Crosses have been added to the appropriate towns, and hundreds are shown.
    (b) c.1693. Four plans occupy the space previously taken by the Royal Arms and Saxton's title and its cartouche, i.e. 'Breknoke', 'Caermarden', 'Cardigan', and 'Radnor'. The small erasure in the bottom right-hand corner has been ruled off to form a small panel, but this remains blank. Further hundreds are shown, and roads and eight heraldic shields added.

27. **Shropshire** (preceding state, 1665)
    No significant changes of format occur, most having been accomplished by 1665, which date has been all but effectively erased. Crosses and crowns have been added to the appropriate towns by c.1689, and roads appear by c.1693.

28. **Somerset** (preceding state, 1665)
    No significant changes of format occur, most having been effected by 1665. By c.1689, the date had been crudely altered from 1665, mitres added to Bath and Wells, and crowns and crosses to the appropriate towns. By c.1693 the title had been re-engraved, and roads were shown.

29. **Staffordshire** (preceding state, 1665)
    No significant changes of format occur, most having been effected by 1665. By c.1689, the map had a new English title replacing the Latin one of 1665, a previously blank shield had the arms of the Stafford family inserted, a mitre had been added for Lichfield, and crosses and crowns to the appropriate towns. By c.1693 the roads had appeared, and the arms of 'Will: Howard Viscount' had been added along the bottom edge of the map.

30. **Suffolk** (preceding state, 1665)
    No significant changes of format occur, most having been effected by 1665. Crowns and crosses had been added to the appropriate towns by c.1689, and by c.1693 the title had been re-engraved, roads added, and the arms of 'Hen. How. E.' inserted.

31. **Warwickshire** (preceding state, Web; 1645)
    (a) c.1689. The Royal Arms of Elizabeth I and Saxton's Latin title, both of which had survived in Web's edition, have been erased, and superseded by a plan of 'Warwicke'. Seckford's arms have been replaced by a plan of

	Coventry, while a plan of 'Lecester' replaces the lower part of Saxton's scale which contained the dedication and reference to Leonard Terwoort. This has involved the re-engraving of the line-scale itself, although Saxton's open dividers, which surmount it, remain unaltered. The new English title occurs between the plans of Warwick and Coventry, and is surmounted by heraldic shields. Hundreds had been added, as too had a mitre for Coventry and crowns for the market-towns of Warwickshire alone.
	(b) *c.*1693. No significant changes in format have occurred, but the roads have been added, Lichfield has a crown and mitre, and crosses and crowns have been inserted in Leicestershire.
32. **Westmorland** (preceding state, Web; 1645)	(a) *c.*1689. Web's date (1642) has been crudely erased and the space left blank. A road 'To Lond.' is shown, and crosses are added to some market-towns. But no significant changes of format had occurred.
	(b) *c.*1693. The title has been re-engraved in the same position as formerly, but below it occur five heraldic shields (two of which, on the extreme right, are blank), and five monumental columns. A plan of 'Carlile' replaces Saxton's scale, which has been re-engraved on the right of the map, but without the open dividers. Wards are shown, a crown and mitre have been added for Carlisle, additional crosses appended to some market-towns, and all the roads of Ogilby's survey have been inserted.
33. **Wiltshire** (preceding state, 1665)	No significant changes of format occur, most having been effected by 1665. By *c.*1689, the date had been crudely altered from 1665, and crowns, crosses, and mitres added to the appropriate towns. By *c.*1693 the roads had appeared, and the arms of 'Edw. Hide' had been completed, in the bottom right-hand corner.
34. **Worcestershire** (preceding state, 1665)	No significant changes of format occur, most having been effected by 1665. By *c.*1689 a mitre had been added to Worcester, and crowns and crosses to the appropriate towns, and some roads had appeared. By *c.*1693 the title had been re-engraved, and the roads were shown more fully, together with some destinations beyond the county boundary, e.g. 'from Hereford' and 'from Aberisthwith'.
35. **Yorkshire** (preceding state, 1665)	(a) *c.*1689. The wapentakes have been named, a mitre, crosses, and crowns added to the appropriate towns, and roads (after Ogilby) added; but no major changes in the format have occurred.
	(b) *c.*1693. The plan of York, which had survived since Web added it in the edition of 1645, has been erased, and replaced by a new English title surrounded by fourteen heraldic shields. The view of Hull, too, together with Saxton's original scale, have been erased, but the spaces have been left blank. A new vertical scale has been added on the left-hand side.

Appendix 14

LATER EDITIONS OF SAXTON'S ATLAS. EXTANT COPIES

William Web, 1645
Cambridge. University Library   Atlas. 4.64.3
London. British Library, Map Room   c.7c.3
Oxford. Bodleian Library   Gough Maps 96*

Philip Lea, c.1689
London. British Library, Map Room   c.21.e.10 (*ex. Gardner; ex. Fordham*)
London. Royal Geographical Society   264.H.17
Oxford. Bodleian Library   Douce Prints b.28

Philip Lea, c.1693
Bangor. University College of North Wales Library   912.42
Blackburn. Public Library (French edition)   912.42 LEA: (see note below)
Cambridge. Magdalene College, Pepysian Library   PL 2987
Cambridge. University Library   Atlas. 4.69.2

Leeds. University of Leeds, Brotherton Library	Whitaker Coll. 2
London. British Library, Map Room	c.7c.4
London. British Library, Map Room	c.21.e.11 (*ex. Baxter*)
London. National Maritime Museum	SAX 002.1
Manchester. Central Reference Library	BR f 912.42 S6
Newcastle-upon-Tyne. City Library	912.42. Book No.D1629
Oxford. Worcester College Library	R.7.10
Washington D.C. (U.S.A.). Library of Congress	G 1807. S3 1690 Map Cage
(private) J. Gardner	—
(private) D. Hodson	—

George Willdey, c.1730

Leeds. University of Leeds, Brotherton Library	Whitaker Coll. 3
Oxford. Bodleian Library	Gough Maps 90

Thomas Jefferys, c.1749

Aberystwyth. National Library of Wales	Cupboard B. Shelf 5, No.21
London. British Library, Map Room	c.21.e.12 (*ex. Baxter*)

C. Dicey & Co., c.1770

Leeds. University of Leeds, Brotherton Library	Whitaker Coll. 4

Lost Copies

Newport (Monmouthshire). Public Library	f.094. Catalogued as of 1690, and as wanting titlepage. Has disappeared

Note The Blackburn copy was reported missing, 1979.

Appendix 15

ACTS OF THE PRIVY COUNCIL
P.R.O. PC 2/10 p.443

At Westminster, the xjth of Marche, 1575

A placart to (blank) Saxton, servant to Mr Sackeford, master of the Requestes, to be assisted in all places where he shall come for the view of mete places to describe certein counties in cartes, being thereunto appointed by her Majesties bill under her signet.

Appendix 16

PATENT ROLL 17 Eliz. Pt.7 (No.1129), P.R.O. C.66/1129
(Abstract from Latin)

The Queen &c. Know all we, in consideration of the services of Christopher Saxton in and about the survey and description of all and singular the counties of England, have given and granted to the same Christopher the office of bailiff, collector and receiver of all the rents and profits of all the manors, messuages, lands etc. of us our heirs ands successors in the city of London and in the county of Middlsex late belonging to the Priory or Hospital of St. John of Jerusalem in England, and parcel of the possessions thereof late existing. To hold the said office to the said Christopher Saxton for himself or his deputy or deputies when the office shall happen to be vacant by the death, forfeiture etc. of Constant Benet now occupying the office, for the life of Christopher together with the wages and fees of 10*l* yearly and all profits, advantages etc. to the same office belonging To be received yearly from the issues profits and revenues of all and singular the aforesaid manors, lands etc as well by his own hand and retained in his own hand as by the hand of the farmer, tenant, or any other person occupying the premises or any of them at four annual terms etc.
Westminster 19 January 17 Elizabeth (1574)

Appendix 17

GRANT OF ARMS TO CHRISTOPHER SAXTON
(Bodleian Library: MS. Ashmole 834 fol.22v)

I Willm̄ Flower Esquire als Norroy King of Arms of the north partes of this realme of England, being required by Christopher Saxton of Dunningley in the countie of York gentleman, to discrybe and delyver unto him the Armes of his ancestors in such sort as he may lawfully use and beare the same, have thought good in respect of the worthynes of the said Christopher Saxton who by speciall direction and commandment from the Queens ma^{ti} hath endevored to mak a perfect geographicall description of all the severall shires and countis within this realme, and accordingly finished the same to his everlastinge prayse, not only to condiscend unto his said request, but also further to the perpetuall remembrance of his well spent tyme therin and in signification of his desert that way, to add unto his ancyent Armes being Argent three chaplets or garlands in bend gules double cotysed sable, for his crest upon the healme on a Torc or Wreath argent and sable, the demy Arme of a man, with the sleeve gold, the hand proper culor holding a payre of compasses gold, therunto ioyning, mantelle of gules duble argent, as more playnly is to be seene depicted in these presentes. Which Armes and crest and every part and parcell thereof I the sayd Norroy King of Arms, do give and graunt, ratify, confirm, delyver and assign unto the sayd Christopher Saxton, and to his posterity for ever by these presents. Signed and subscribed with myne owne hand the first day of July Anno Dni. 1579 in the xxjth yere of the reigne of our most gracious soueraigne Lady Queen Elizabeth.

Appendix 18

PEDIGREE OF THE SAXTON FAMILY BEARING THE ARMS

Argent on a Bend Gules Double Cottised Sable, Three Chaplets or Garlands. From *Visitations of Yorkshire 1584–5 and 1612*, Edited by Joseph Foster, 1875.

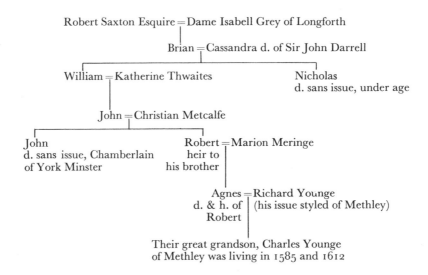

A slightly different version of the above is found in the Hailstone MS (York Minster Library), compiled by John Hopkinson (MS.Add. 164/1 p.304) in which Brian Saxton had a brother Robert, who had issue Nicholas, who had issue Nicholas. Christopher Saxton, the surveyor, may have been descended from this line.

The Percehay family of Barton on Humber also bore the Saxton Arms as a quartering, acquired on the marriage of Walter, or William, Percehay of Ryton to Mary Saxton whose pedigree is shown below. From *The Visitation of Yorkshire in the Years 1563 & 1564, Made by William Flower Esq. Norroy King of Arms*, Edited by C. B. Norcliffe, 1881.

```
Sir Hugh Saxton of Prescot in Flyntshire =
│
Robert =
│
Nycolas =
│
William =
├─────────────────┐
William =        Bryan =
│
Robert Esquyer for the Body to Edward 4
│
Robert Merchant Taylor of London = ... d of ... Kebell
│
Mary = W. Percehay of Ryton
├─────────────────────────┐
Richard Percehay         Antony Percehay
s. & h. Bat. of          2nd son
Syvell Lawe,
Christchurch,
Oxford
```

Appendix 19

PATENT ROLL 22 Eliz. Pt.2 (No.1187), P.R.O. C.66/1187
(Abstract from Latin)

The Queen etc. Know all that we for divers causes and considerations have sold, granted and to farm demised to Christopher Saxton all that our parcel of waste land lying in the street or by the street called 'St John Strete' in co. Midd. within the parish of St. Sepulchre without Newgate London and without the bars of West Smithefeld in the City of London, to wit between the Kings road going to Iselington alias Iseldon unto Westmithfeld unto the street called Turmill Street on the west side and abutting upon another little way going to the said street called Turmill street unto a street called St. John's street towards the north and upon the parcel of waste aforesaid usually for the way to Turmill Street aforesaid towards the south which parcel of waste above granted contains in length north and east, to wit, north-east 16 virgates and 2 feet and contains in length west and south, to wit, south-west 20 virgates and contains in breadth at one end of the same abutting towards the north 16 virgates and contains in breadth at the other end of the same waste abutting towards the south 4 virgates and 2 feet which same parcel of waste land is or was parcell of the manor of Clarkenwell and late of the monastery or priory of Clarkenwell now dissolved or of the priory of St. John of Jerusalem in England late dissolved formerly parcel of the possessions thereof. To hold the said waste ground to Christofer Saxton his executors and assigns from the feast of the Annunciation last past for the term of 60 years, rendering yearly to us our heirs and successors 3s 4d yearly during the term aforesaid, with permission for Christofer Saxton his executors or assigns to build one or more houses upon the said piece of waste ground.
Westminster 1 July 22 Elizabeth (1580)

Appendix 20

ACTS OF THE PRIVY COUNCIL
(Eliz. Vol.5 p.79)

At Nonesoche, the xj of Julie 1580

An open letter to all Justices of the Peace, Maiours, Sherifes, &c., that where her majestie hathe graunted unto Christofer Saxton a lease of a peece of grounde without the Barres of Westmithfield in the countie of Middlesex, to the intent to build certen convenient buildings there, as, namelie, a Sessions Howse for the Justices in Middlesex to kepe Sessions of Enquirie for that countie, Their Lordships do signifie that notwithstanding the late Proclamacion to restraine buildinges here about the Cittie, it is not ment that her

Majesties graunt to Saxton for the building as before shold anie waie be impeached, but suffered to proceade according to the said graunt, their Lordships, in her Majesties name, do will and comaund every of them to whom it maie appertaine quietly to suffer the said Christofer and his workmen to fynishe the same, notwithstanding the Proclamacion.

Appendix 21

PATENT ROLL 26 Eliz. P.R.O. C.66/1239
(Abstract from Latin)

The Queen &c. Know ye that for various good causes and considerations, prompted by the advice of William, Baron of Burghley, Treasurer of England, and of Walter Mildmaye, Knight, Chancellor of our Court of Exchequer, we have conveyed, granted and leased at farm, to our beloved Christofer Saxton, gentleman, all that our Rectory and Church of Scawbye with all its rights, in the county of York, now or formerly in the tenure or occupation of William Clifton and John Singleton or their assigns, formerly belonging to the monastery or priory of Bridlington in the county of York; and also all and singular the houses, buildings, lands commons &c; tithes of sheaves, (standing) corn, grain, hay, wool &c., and all other tithes, both great and small, as well as our offering, revenues, profits, emoluments &c., whatsoever in any way belonging to the Rectory or Church of Scawbie for or under the annual rent specified below; except, however, all the great trees, woods, underwoods, mines and quarries, and the advowson of the vicarage of the parish church of Scawbie aforesaid; Christofer Saxton his heirs & assigns to have and to hold the aforesaid rectory, lands, tenements, meadows, glebe lands, tithes of sheaves etc. leased by this deed, except as stated, from Lady Day last past until the end of the term, and for the term, of twenty-one years henceforth following;

Rendering therefor annually to us ten pounds and ten shillings at Michaelmas and Lady Day into the hand of the Bailiff or Receiver, to be paid by equal portions during the term. And Christofer Saxton, his executors and assigns shall sufficiently repair and maintain both the chancel of the church aforesaid, as well as the houses and buildings, and all hedges, ditches, enclosures, shores, banks and seawalls, as often as may be necessary at his or their own expense.

And we will and grant to Christofer Saxton &c. that they be allowed to take sufficient houseboote, hedgeboote, fyreboote, ploughboote and carteboote there, and that they shall have the timber growing in the woods towards the repair of the chancel and houses under the direction of the Steward or Understeward. Provided that neither Christofer Saxton &c. shall be charged with the rent until Lady Day 1588, unless the Rectory and the rest of the things granted shall have gone into the possession of Christofer Saxton, by virtue of these our Letters Patent; Provided also that if, at any time after Lady Day 1588, the rent shall become in arrear by the space of forty days, then this lease and grant shall be void.

Witnessed by the Queen at Westminster 27th June 1584. By commission of warrant.

Appendix 22

LETTERS PATENT: DUCHY OF LANCASTER P.R.O. DL 13/6 Pt.2
(Abstract from Latin)

3 November 31 Eliz. (1589)

Letters Patent issued under the seal of the Duchy of Lancaster, reciting that by Letters Patent issued under the same seal dated 26 April 22 Elizabeth (1580), John Foxcrofte was appointed to the office of bailiff of West Pontefract and the wapentakes of Agbrigg and Morley, part of the Duchy of Lancaster in the county of York, to occupy to himself or his sufficient deputy or deputies from the date of the Letters Patent during the Queen's pleasure, with the wages, fees and profits of the office annually due and customary, receiving annually the same wages and fees proceeding and arising from the issues, profits and revenues of the premises by the hands of the receiver of the Duchy at Easter and Michaelmas, together with all other profits, rewards, advantages and emoluments lawfully belonging and by right pertaining to the same office in as full a manner and form as one Edmund Wilbore or any other having, executing or occupying the said office before, and lastly doing as fully as is contained in the Letters Patent. And we do terminate this appointment for certain reasons specially moving us, and by this annullment the office of bailiff of West Pontefract and of the wapentakes of Agbrigg and Morley now stands disposable. Through the advice and consent of the council of the Duchy our well beloved Christofer Saxton, gentleman, is appointed to the office or bailiff of West Pontefract and the wapentakes of Agbrigg and Morley, parcel of the Duchy of Lancaster, to occupy the office for himself or his sufficient deputy or deputies from the date hereof during the Queen's pleasure, with the wages, fees and profits anciently due and customary, to receive the wages, fees and revenues from the hands of the receiver of the Duchy at Easter and Michaelmas in even portions, together with all other profits, rewards, advantages and emoluments whatever lawfully belonging and by right pertaining to the office in as full a manner as Edmund Wilbore and John Foxcrofte or any other having, executing or occupying the said office before, providing that Christofer Saxton or his deputy or deputies do duty annually at the county assizes and sessions as is usual for the bailiff, and providing also that Christofer Saxton or his deputy or deputies renders a true account annually, before the auditor of the Duchy of Lancaster when he comes into those parts, of the issues, profits and revenues issuing from the said office of all that pertains to the Queen, and her heirs, and provided that these Letters Patent are enrolled within one year before the auditor, or they will be void. Given &c. the third day of November in the 31st year of our reign.

.... Walsyngham
.... Brograve
pp Will: Gerrard

Appendix 23

LETTERS PATENT: DUCHY OF LANCASTER P.R.O. DL 13/6 Pt.2

Wheras may appear unto you by a particular rated by me for that purpose the xviijth daie of October last a smale Balyweek or twoe in Yorkshire are to be passed in patent to Christopher Saxton gent. Theise are to will and require you to take bonds of him the said Christopher in the some of fiftie pounds to hir Maties use that he shall truly dischardge himselfe & be from time to time accounptable for all things concerning his said office & in the said bond that there be bownd wth him as surties the persons undernamed & if anie of them happen not to be now in the Citie that then you direct out dedimus poestatem for their bonds to be taken in the Countrie. And in so doing these shalbe yor warrant. At the Savoy the   of November 1589
Fra: Walsyngham

                                            Robert Mawe of Lincolns Inne gent
                                            Peter Page of the countie of   gent
                                            Thomas Hall of the countie of   gent
                                            or anie twoe of them

To my loving frend Mr Willm̄ Garrard clerke of the Duchie

Appendix 24

Extract from: *Select Tracts and Tables Relating to English Weights & Measures* (1100–1742)
Ed. Hall and Nicholas, Camden Miscellany Vol.XV p.28

Calendar & Tables (MS. Harl. 5769) 1682
An Inch is 3 barley cornes dry & round in length. A Foote is 12 inches. A Cubit is a foote and half. A Yard is two Cubits or three feete. An Elle is a yard & 9 inches. A Fadome is two yards. A Pearch, or a Rod, or a Pole (by statut) must be 5 yards and a half; or 16 feete and an half. But in some places of England they measure wth a pearch of 12 foote called Tenant right or Court measure. In other places they measure wth a pearch of 18, 20, or 24 foote, called Woodland Measure. A Score is 20 yards. A Furlong is 40 pearches in length. A Myle is 8 furlongs, or 320 pearches. An Acre is one pearch in breadth and 160 in length; so 4 in breadth and 40 in length; and so according to that proportion. A Daywork is 1 pearch in breadth and 4 in length. A Roode or a Farthendale conteynes 10 day workes; that is, one pearch in breadth and 40 in length. A yard is 4 acres. A Hyde of land is five yards of land. A Knightes fee conteynes 8 hydes of land, which is plough-till a year.

References

CHAPTER 1
1. Borthwick Institute of Historical Research: Prob.Ref.23 Fol.634A.
2. *Ibid*. Prob.Ref.11 Fol.113.
3. *Ibid*. Prob.Ref.13 Fol.394v.
4. Hunter, Joseph. Birkbeck-Nalson pedigree, Familae Minorum Gentium Vol.2: *Harleian Society* 38 (1895) 822.
5. The entry in the Dewsbury Parish Register of the marriage of Robert Saxton o.t.p. and Ann Oxeley of Hoyland, 10.2.1606/7, does not refer to the same Robert Saxton as here discussed.
6. Yorkshire Archaeological Society: DD.146 ix.
7. Wakefield M.D.C. Archives, Goodchild Loan: Charlesworth deeds.
8. Field, Osgood. 'John Field of East Ardsley, "The Proto-Copernican of England"' *Yorkshire Archaeological Journal* (hereafter Y.A.J.) 14 (1898), 81–4.
9. Walker, J. W. *Wakefield its History and People*, 2nd ed. (1939), 367.
10. Cooper, Charles H. and Thompson. *Athenae Cantabrigiensis Vol 1, 1500–1588* (1858).
11. Marcombe, David. 'Saxton's Apprenticeship: John Rudd, A Yorkshire Cartographer', *Y.A.J.* 50 (1978), 171–5.
12. Durham Chapter Records: Treasurer's Book No.7 (1569–70), Stipends.
13. Bodleian Library: MS. Wood D.13.
14. Fordham, Sir H. George, Additional Note, *Thoresby Society* 28 (1927), 491.

CHAPTER 2
1. Andrews, John H. 'Christopher Saxton and Belfast Lough'. *Irish Geography* 5 (1965), 1–6.
2. Hind, Arthur M. *Engraving in England in the Sixteenth and Seventeenth Centuries. A Descriptive Catalogue with Introductions. Part 1. The Tudor Period* (1952).
3. Lynam, Edward. *Saxton's Atlas of England and Wales. An Introduction* (1936).
4. *Ibid*.
5. Manley, Gordon. 'Saxton's Survey of Northern England'. *Geographical Journal* 83 (1934), 308–16.
6. Skelton, Raleigh A. *County Atlases of the British Isles, 1579–1850. A Bibliography* (1970).
7. Hind. *op. cit.*
8. Manley. *op. cit.*
9. *Ibid*.
10. *Ibid*.

CHAPTER 3
1. Chubb, Thomas. *The Printed Maps in the Atlases of Great Britain and Ireland. A Bibliography, 1579–1870* (1927).
2. Hind, Arthur M. 'Queen Elizabeth by Augustine Ryther (?): An Undescribed State'. *British Museum Quarterly* 9 (1934–5), 58.
3. Skelton, Raleigh A. *County Atlases of the British Isles, 1579–1850. A Bibliography* (1970).
4. Hind, Arthur M. *Engraving in England in the Sixteenth and Seventeenth Centuries. A Descriptive Catalogue with Introductions. Part 1. The Tudor Period* (1952).
5. Laxton, P. *Two Hundred and Fifty Years of Map-making in the County of Hampshire. A Collection of Reproductions of Printed Maps Published Between the Years 1575 and 1826* (1976).
6. Heawood, Edward. 'The Use of Watermarks in Dating Old Maps and Documents'. *Geographical Journal* 63 (1924), 391–412.
7. Hind. *op. cit.* (1934–5).
8. Skelton. *op. cit.*
9. Seebohm, Frederic. *Customary Acres and their Historical Importance* (1914); Close, Sir Charles. 'The Old English Mile', *Geographical Journal* 76 (1930), 338–42; Karslake, J. B. P. 'Further Notes on the Old English Mile', *Geographical Journal* 77 (1931), 358–60; Evans, Ifor M. 'A Cartographic Evaluation of the Old English Mile', *Geographical Journal* 141 (1975), 259–64.
10. Evans. *Ibid*.
11. Manley, Gordon. 'Saxton's Survey of Northern England'. *Geographical Journal* 83 (1934), 308–16.
12. Skelton, Raleigh A. *Saxton's Survey of England and Wales: With a Facsimile of Saxton's Wall-map of 1583* (1974).
13. Thoresby, Ralph. *Ducatus Leodiensis, or the Topography of the Town and Parish of Leeds and Parts Adjacent in the West-Riding of York* (1715).
14. Hind. *op. cit.* (1952).
15. *Ibid*.
16. Skelton. *op. cit* (1970).

17. Lynam, Edward. *Saxton's Atlas of England and Wales. An Introduction* (1936).
18. Walpole, Horace. *Anecdotes of Printing in England* (1786).
19. Hind. *op. cit.* (1952).

Chapter 4

1. Crone, Gerald R. *et al.* 'Landmarks in British Cartography'. *Geographical Journal* 128 (1962), 406–30.
2. Marcombe, David. 'Saxton's Apprenticeship: John Rudd, a Yorkshire Cartographer'. *Y.A.J.* 50 (1978), 171–5.
3. Lynam, Edward. 'English Maps and Map-makers of the Sixteenth Century'. *Geographical Journal* 116 (1950), 7–28.
4. Manley, Gordon. 'Saxton's Survey of Northern England'. *Geographical Journal* 83 (1934), 308–16.
5. Bedford, W. K. R. 'The Weston Tapestry Maps'. *Geographical Journal* 9 (1897), 210–15.
6. North, F. J. 'The Map of Wales'. *Archaeologia Cambrensis* 90 (1935), 1–69.
7. Fordham, Sir H. George. 'Christopher Saxton of Dunningley: His Life and Work'. *Thoresby Society* 28 (1927), 356–84.
8. Darby, Henry C. 'The Agrarian Contribution to Surveying in England'. *Geographical Journal* 82 (1933), 529–35.
9. Richeson, A. W. *English Land Measuring to 1800: Instruments and Practices* (1966).
10. Taylor, Eva G. R. 'The Plane-table in the Sixteenth Century'. *Scottish Geographical Magazine* 45 (1929), 205–11.
11. Foullon, Abel. *L'Holometre* (1551).
12. Worsop, Edward. *A Discoverie of Sundrie Errours and Faults Daily Committed by Landmeaters* (1582).
13. Digges, Leonard. *A Book named Tectonicon* (1556).
14. Bourne, William. *A Treasure for Travellers* (1578).
15. Digges, Leonard. *A Geometrical Treatise named Pantometria* (1571).
16. Taylor, Eva G. R. 'The Earliest Account of Triangulation'. *Scottish Geographical Magazine* 43 (1927), 341–5.
17. Digges. *op. cit.* (1556).
18. *Ibid.*
19. Digges. *op. cit.* (1571).
20. Leigh, Valentine. *The Most Profitable and Commendable Science of Surveying of Landes, Tenemens, and Hereditamentes* (1577).
21. Lucar, Cyprian. *A Treatise Named Lucar Solace* (1590).
22. Agas, Ralph. *A Preparative to Plotting of Lands and Tenementes for Surveighs* (1596).
23. Rathborne, Aaron. *The Svrveyor in Foure Bookes* (1616).
24. Leybourn, William. *The Compleat Surveyor, Containing the Whole Art of Surveying of Land. By the Plaine Table, Theodolite, Circumferentor, Peractor, and other Instruments* (1653).
25. Manley, Gordon. 'Observations on the Early Cartography of the English Hills'. *Comptes Rendus du Congrès International de Géographie* 2 (1938), 36–42.
26. Evans, Ifor M. 'A Cartographic Evaluation of the Old English Mile'. *Geographical Journal* 141 (1975), 259–64.

Chapter 5

1. Lynam, Edward. *Saxton's Atlas of England and Wales. An Introduction* (1936).
2. Whitaker, Harold. 'The Later Editions of Saxton's County Maps'. *Imago Mundi* 3 (1939), 72–86.
3. Chubb, Thomas. *A Descriptive List of the Printed Maps of Somersetshire, 1575–1914* (1914).
4. Whitaker, Harold. *A Descriptive List of the Printed Maps of Lancashire, 1577–1900* (1938); *A Descriptive List of the Printed Maps of Northamptonshire, 1576–1900* (1948).
5. Whitaker, Harold. *A Descriptive List of the Printed Maps of Yorkshire and its Ridings, 1577–1900* (1933).
6. Whitaker, Harold. *A Descriptive List of the Printed Maps of Cheshire, 1577–1900* (1942); *A Descriptive List of the Maps of Northumberland, 1576–1900* (1949).
7. Chubb. *op. cit.*
8. Chubb, Thomas. *The Printed Maps in the Atlases of Great Britain and Ireland. A Bibliography, 1579–1870* (1927).
9. Fordham, Sir H. George. *Hertfordshire Maps. A Descriptive Catalogue of the Maps of the County, 1579–1900* (1907); *Cambridgeshire Maps: A Descriptive Catalogue of the Maps of the County and of the Great Level of the Fens, 1579–1900* (1908).
10. Curwen, John F. *The Chorography, or A Descriptive Catalogue of the Printed Maps, of Cumberland and Westmorland* (1918).
11. Whitaker. *op. cit.* (1939).
12. *Ibid.*
13. Ogilby, John. *Britannia* (1675).
14. Tyacke Sarah. *London Mapsellers, 1660–1720* (1978).

CHAPTER 6
1. Copinger, W. A. *The Manors of Suffolk*, 5 (1905–11), 174–6.
2. Stow, John. *The Survey of London*, first pub. 1598.
3. Public Record Office (hereafter P.R.O.), DL.28/10/1, and Somerville, Sir Robert, *History of the Duchy of Lancaster, Vol.1, 1265–1603* (1953), 330.
4. Bodleian Library, MS. Ashmole 834 fol.22v. The copy is MS. Ashmole 858 fol.34.
5. Norfolk Record Office: Dean & Chapter of Norwich, Leases.
6. P.R.O. DL.13/6 Pt.2.
7. P.R.O. DL.42/99 ff.68v–69.
8. Somerville, Sir Robert. *Office Holders in the Duchy and County Palatine of Lancaster From 1603* (1972).
9. Trust Deed at West Yorkshire County Record Office.
10. Robertshaw, Wilfred. 'Notes on Adwalton Fair', *The Bradford Antiquary* 7 (1927), 51; and 'New Light on Adwalton Fair', *The Bradford Antiquary* 10 (1962), 279.
11. Patent Roll (Chancery) 19 Eliz. Pt.6 in (37) 29, 1577 (Quoted in Parker, James, *Illustrated History From Hipperholme to Tong* (1904), 301.

CHAPTER 7
1. Fordham, Sir H. George. 'Christopher Saxton, of Dunningley. His Life and Work', *Thoresby Society* 28 (1927), 356–84.
2. Evans, Ifor M. 'A Newly Discovered Manuscript Estate Plan by Christopher Saxton', *Geographical Journal* 138 (1972), 480.
3. Fisher, J. L. *Essex Archaeological Soc.*, 23 (1924–5), Appendix 3, 92–6.
4. Zupko, Ronald E. *A Dictionary of English Weights and Measures From Anglo Saxon Times to the Nineteenth Century* (1968), 120–22.

CHAPTER 8
1. P.R.O. E.178/2690.
2. Tempest, Eleanor B. 'Broughton Hall and its Associates', *The Bradford Antiquary*, 6 (1921), 83.
3. P.R.O. E.178/2694.
4. Greater London Record Office: H1/ST/A1/4.
5. *Ibid*. H1/ST/E29/2.
6. Hasted, E. *The History and Topographical Survey of the County of Kent* (1797, reprinted 1972).
7. Folger Shakespeare Library, Washington D.C. MS. 135.4
8. Greater London Record Office: H1/ST/E106.
9. *Ibid*. H1/ST/E115/34.
10. Essex Record Office, D/D Th 18.
11. Eden, Peter (Ed.). *Dictionary of Land Surveyors and Local Cartographers of Great Britain and Ireland 1550–1850* (1976), F 209.
12. P.R.O. SP.46/18 222d, 223.
13. *Ibid*., E.178/2713.
14. British Museum, Additional MS. 50189, and Harvey, P. D. A. 'A Manuscript Map by Christopher Saxton', *The British Museum Quarterly*, 23 No.3 (1961).
15. Kent County Archives Office U 390 P2, and Evans, I. M. 'A Newly Discovered Estate Plan by Christopher Saxton, Relating to Faversham in Kent', *Geographical Journal*, 138 Pt.4 (1972), 480–86.
16. P.R.O. SC. 12/1/1.
17. *The Bradford Antiquary* 9 (1940), 1.
18. P.R.O. MPB.30.
19. Dodd, E. E. 'Priestthorpe and the Rectory of Bingley', *The Bradford Antiquary*, 10 (1952), 1.
20. P.R.O. E.178/2746.
21. Speight, Harry. *Chronicles of Old Bingley* (1899).
22. Darbyshire, Hubert and Lumb, George D. 'The History of Methley', *Thoresby Society* 35 (1937), 118–30.
23. Borthwick Institute, HC.AB 12 ff 98v–100v.
    Note: Darbyshire and Lumb transcribe texts concerning the decrees (pps.40–3 above), but should the reader consult this source they should be aware that the date given on the first line of the second decree, quoted as 1592, is incorrect and should read 1582.
24. Suffolk Record Office JA2/7/1.
25. Copinger, W. A. *The Manors of Suffolk* 2 (1905–11), 14.
26. Nottinghamshire Record Office DD. SR 202/32.
27. *Ibid*. DD. SR 202/34.
28. Hunter, Joseph. *South Yorkshire* Vol.2 (1831, reprinted 1974), 302.
29. Nottinghamshire Record Office DD. SR 227/62.
30. Whitaker, Thomas D. *Loidis and Elmete* (1816), 312.
31. Nottinghamshire Record Office DD. SR 227/62.
32. *Ibid*. DD. SR 202/50.
33. *Ibid*. DD. SR 227/62.

34. *Ibid*. DD. SR 1/18/5/1.
35. *Ibid*. DD. SR 1/18/5/2.
36. Dunstan, George. *The Rivers of Axholme* (nd.), 114–9.
37. P.R.O. MPB 16.
38. P.R.O. E.134 (39 Eliz.) E.14.
39. 'The Private Diary of Dr John Dee', *Camden Society* (1842), 55–6.
40. Pierpont Morgan Library, New York. See Seymour de Ricci and W. J. Wilson, *Census of Medieval & Renaissance Manuscripts in the United States and Canada* Vol.2 (1937), 1626.
41. Salop Record Office, Bridgewater Collection: SR.212/Box 346(9).
42. Hopkins, Eric. *The Bridgewater Estates in North Shropshire in the First Half of the Seventeenth Century*, MA Thesis, London 1956 (unpublished).
43. P.R.O. MPC 9.
44. The author is grateful to the headmaster of Horbury School for allowing inspection of copies of the original documents, taken by John Charlesworth in 1898, now at the school.
45. P.R.O. MPB 32.
46. Beresford, Maurice. 'A Journey Along Boundaries', *History on the Ground* (1957), 52–6; and P.R.O. E.132/24 ff 167 and 283.
47. P.R.O. MPC 111.
48. P.R.O. DL 4/44/47.
49. Halifax Antiquarian Society's Papers (1923), 21–63.
50. Brotherton Library, The University of Leeds: MS. (Deposit) 1949/1/25.
51. *Ibid*. MS. (Deposit) 1946/1/8/6.
52. Wentworth, George E. 'The Wentworths of Woolley', *Y.A.J.* 12 (1893), 5–6.
53. Goodricke, Charles A. *History of the Goodricke Family* (1885), 12–24.
54. Yorkshire Archaeological Society, MS. 930b and 930c.
55. Leeds City Archives, Harewood Deeds 221.
56. Speight, Harry. *Kirkby Overblow and District* (1903), 87–91; and Goodricke, Charles A., *op. cit*.
57. Chatsworth House, Cavendish Archives.
58. Wakefield M.D.C. Archives, Goodchild Loan.
59. Durant, David N. *Bess of Hardwick* (1977), 238; (Hardwick Drawer 337).
60. Walker, J. B. *Wakefield its History and People*, 2nd edition (1939), 661.
61. Pilkington, Lt. Col. J. *The History of the Pilkington Family of Liverpool* (1894), 5–7.
62. Yorkshire Archaeological Society MD. 59/19.
63. Fordham, Sir H. George. 'Christopher Saxton of Dunningley. His Life and Work'. *Thoresby Society* 28 Pt.4 (1927), 381–4.
64. North Yorkshire Record Office ZDS.M.5/1.
65. Eden, Peter (Editor). *A Dictionary of Land Surveyors and Local Cartographers of Great Britain and Ireland 1550–1850*, Vol.2 (1976), P.169.
66. *Victoria County History of Rutland*, Vol.2.
67. Alnwick Castle: Syon House MS. 11.6.34.
68. Batho, G. R. 'Two Newly Discovered Manuscript Maps by Christopher Saxton', *Geographical Journal* 125 (1959), 70–5.
69. Quoted by Batho from Harrison G.B. (Editor) *Advice to his Son* (1930), 75, 77.
70. Alnwick Castle X116 32f.
71. P.R.O. MPC 112.
72. P.R.O. DL 4/51/59.
73. Peel, Frank. *Spen Valley Past and Present* (1893), 39, 123–33.
74. Lynam, Edward. 'English Maps and Map-Makers of the Sixteenth Century', *Geographical Journal* 116 (1950), 7–28 (reprinted as Chap.3 of Lynam, E. *The Mapmaker's Art* 1953)).
75. P.R.O. MPF 77.
76. *Belfast Telegraph*, March 24th, 1949.
77. Andrews, John H. 'Christopher Saxton and Belfast Lough', *Irish Geography* 5, Pt.2 (1965), 1–7.
78. Wentworth, George E. 'The Wentworths of Woolley', *Y.A.J.* 12 (1893), 6.
79. P.R.O. DL. 4/9/46.

CHAPTER 9
1. Pierpont Morgan Library, New York. (See reference Chapter 8/40.)
2. Leeds City Archives DB.213/23.
3. Domestic – Addenda Vol.23. Quoted in full in *Chapters of Yorkshire History* James J. Charlesworth (1872), 73–6.
4. Leeds City Archives TN/WK/L/3.
5. Yorkshire Archaeological Society DD.12 Add.
6. Whitaker, Thomas D. *Loidis and Elmete* (1816), 219–29.
7. Yorkshire Archaeological Society DD.12: parcels 36–38.
8. Walker, E. J. 'Our Local Portfolio' No.143, *Halifax Guardian* February 25th, 1860.

9. Upton, Anthony F. *Sir Arthur Ingram c.1565–1642. A Study of the Origins of an English Landed Family* (1961).
10. Leeds City Archives TN/AR.
11. Yorkshire Archaeological Society DD 146/11/2.
12. Baildon, W. Paley. *Baildon and the Baildons* (1913), 538–9.
13. Cliffe, J. T. *The Yorkshire Gentry From the Reformation to the Civil War* (1969), 58.
14. Whone, Clifford. 'Christopher Danby of Masham and Farnley', *Thoresby Society* 37 (1936), 1–29.
15. P.R.O. Map: MPC 46; Survey: MPC 65.
16. Somerville, Sir Robert. *Office-Holders in the Duchy and County Palatine of Lancaster From 1603* (1972), 75.
17. Yorkshire Archaeological Society DD.12/1: parcel 27.
18. Leeds City Archives: Stansfield 854.
19. Slater, Philemon. *History of the Ancient Parish of Guiseley* (1880), 83–93.
20. P.R.O. Map: MPC 210; Survey DL.44/953.
21. Robertshaw, Wilfred. 'The Township of Manningham in the Seventeenth Century', *The Bradford Antiquary* 8 (1940), 57.
22. P.R.O. DL.4/61/64.
23. Peel, Frank. *Spen Valley Past and Present* (1893), 146–8.
24. P.R.O. Map: MR 408; Proceedings: E.178/4865.
25. Leeds City Archives LD/170.
26. *Ibid*. Badgery 1056/481.
27. Sheard, Michael. *The Records of Batley* (1896).
28. Leeds City Archives: Badgery 132/1 (1056/459).
29. P.R.O. MPC 53.
30. Bradford Central Library, Local History Library WYC.1617.SAX.
31. P.R.O. DL.4/67/60.
32. Briggs, J. J. 'A Disputed County Boundary', *The Bradford Antiquary* 8 (1940), 1.
33. Willan, T. S. 'Three Seventeenth Century Surveys', *Yorkshire Archaeological Society Record Series* 104 (1941).
34. P.R.O. E.178/4801.
35. Yorkshire Archaeological Society DD.70: bundle 9:11.

Bibliography

Acts of the Privy Council.
Agas, Ralph. *A Preparative to Plotting of Lands and Tenementes for Surveighs* (1596).
Andrews, John H. 'Christopher Saxton and Belfast Lough', *Irish Geography* 5, Dublin (1965), 1–7.
Bagrow, Leo, and Skelton, Raleigh A. *History of Cartography* (1964).
Baildon, W. Paley. *Baildon and the Baildons* (1912–27).
Banks, W. S. *Walks in Yorkshire: Wakefield and its Neighbourhood*. Wakefield (1871).
Batho, G. R. 'Two Newly Discovered Maps by Christopher Saxton', *Geographical Journal* 125 (1959), 70–4.
Bedford, W. K. R. 'The Weston Tapestry Maps', *Geographical Journal* 9 (1897), 210–15.
Beresford, Maurice. 'A Journey along Boundaries', *History on the Ground* (1957), 52–6.
Bourne, William. *A Treasure for Travellers* (1578).
Briggs, J. J. 'A Disputed County Boundary', *The Bradford Antiquary* 8 (1940), 1–16.
Briscoe, A. Daley. *A Tudor Worthy. Thomas Seckford of Woodbridge*. Ipswich (1979).
British Museum. *Catalogue of the Manuscript Maps, Charts & Plans and of the Topographical Drawings in the British Museum* (1962).
Brown, Lloyd A. *The Story of Maps* (1951).
Calendar of Patent Rolls.
Calendar of State Papers, Domestic Series.
Cartwright, James J. *Chapters in the History of Yorkshire*. Wakefield (1872).
Chadwick, S. J. 'Notes on Dewsbury Church and some of its Rectors and Vicars', *Yorkshire Archaeological Journal* (hereafter *Y.A.J.*) 20 (1909), 371–446.
Chadwick, S. J. 'The Dewsbury Moot Hall', *Y.A.J.* 21 (1911), 345–51.
Chubb, Thomas. *A Descriptive List of the Printed Maps of Somersetshire, 1575–1914*. Taunton (1914).
Chubb, Thomas. *The Printed Maps in the Atlases of Great Britain and Ireland. A Bibliography, 1579–1870* (1927).
Clay, J. W. 'The Savile Family', *Y.A.J.* 25 (1920), 1–47.
Cliffe, J. T. *The Yorkshire Gentry from the Reformation to the Civil War* (1969).
Close, Sir Charles. 'The Old English Mile', *Geographical Journal* 76 (1930), 338–42.
Colvin, Sidney. *Early Engraving & Engravers in England (1545–1695), a Critical and Historical Essay* (1905).
Cooper, Charles H. and Thompson. *Athenae Cantabrigiensis Vol.1 1500–1585*. Cambridge (1858).
Crone, Gerald R. et al. 'Landmarks in British Cartography'. *Geographical Journal* 128 (1962), 406–30.
Crone, Gerald R. *Maps and their Makers* (1966).
Crump, W. B. 'A New Saxton Estate Plan', *Y.A.J.* 34 (1939), 359–60.
Crump, W. B. 'Christopher Saxton's Date of Birth', *Geographical Journal* 112 (1948), 256.
Curwen, John F. *The Chorography, or a Descriptive Catalogue of the Printed Maps, of Cumberland and Westmorland*. Kendal (1918).
Darby, Henry C. 'The Agrarian Contribution to Surveying in England', *Geographical Journal* 82 (1933), 529–35.
Darbyshire, Hubert S. and Lumb, George D. 'The History of Methley', *Thoresby Society* 35 (1937).
Davis, R. H. 'Saxton Manuscripts', *Geographical Journal* 139 (1973), 386–7.
Dee, John. 'The Private Diary of Dr John Dee', *Camden Society* (1842).
Digges, Leonard. *A Book named Tectonicon* (1556).
Digges, Leonard. *A Geometrical Treatise named Pantometria* (1571).
Dodd, E. E. 'Priestthorpe and the Rectory of Bingley', *The Bradford Antiquary* 10 (1952), 1–18.
Dunston, George. *The Rivers of Axholme with a History of the Navigable Rivers and Canals of the District*, n.d.
Durant, David N. *Bess of Hardwick* (1977).
Eden, Peter (Ed.). *Dictionary of Land Surveyors and Local Cartographers of Great Britain and Ireland, 1550–1850* (1975).
Ellis, Martha J. 'A Study in the Manorial History of Halifax Parish in the Sixteenth and early Seventeenth Centuries', *Y.A.J.* 49 (1960–1), 250–64, 420–42.
Emmison, F. G. *Catalogue of Maps in the Essex Record Office 1566–1855*. Chelmsford (1947).
Evans, Ifor M. *Christopher Saxton's Atlas of England and Wales 1579 and its Subsequent Editions*. M.A. thesis, Manchester (1971) unpublished.
Evans, Ifor M. 'A Newly Discovered Manuscript Estate Plan by Christopher Saxton, Relating to Faversham in Kent', *Geographical Journal* 138 (1972), 480–6.
Evans, Ifor M. 'A Cartographic Evaluation of the Old English Mile', *Geographical Journal* 141 (1975), 259–64.
Foullon, Abel. *L'Holmetre*. Paris (1551).
Fordham, Sir H. George. *Hertfordshire Maps. A Descriptive Catalogue of the Maps of the County, 1579–1900*. Hertford (1907).
Fordham, Sir H. George. *Cambridgeshire Maps: a Descriptive Catalogue of the Maps of the County and of the Great Level of the Fens, 1579–1900*. Cambridge (1908).
Fordham, Sir H. George. *Studies in Carto-bibliography: British and French, and in the Bibliography of Itineraries and Road-books*. Oxford (1914).

Fordham, Sir H. George. 'Saxton's General Map of England and Wales', *Geographical Journal* (1926).
Fordham, Sir H. George. 'Christopher Saxton of Dunningley: His Life and Work', *Thoresby Society* 28 (1927), 356–84, 491.
Fordham, Sir H. George. 'Some Surveys and Maps of the Elizabethan Period Remaining in Manuscript', *Geographical Journal* 71 (1928), 50–60.
Fordham, Sir H. George. *Some Notable Surveyors and Map-makers of the Sixteenth, Seventeenth, and Eighteenth Centuries and their Work*. Cambridge (1929).
Francois, Martha Ellis, 'The Social and Economic Development of Halifax 1558–1640', *Proceedings of the Leeds Philosophical & Literary Society* 11 (1966), 217–280.
Goodricke, Charles A. *History of the Goodricke Family* (1885).
Greenwood, John B. *The Early Ecclesiastical History of Dewsbury*. Dewsbury (1859).
Hanson, T. W. 'Dr Favour as Protestant Disputant', *Halifax Antiquarian Society's Papers* (1910).
Harvey, Paul D. A. 'A Manuscript Estate Map by Christopher Saxton', *Geographical Journal* 125 (1959), 70–4.
Heawood, Edward. 'The Use of Watermarks in Dating Old Maps and Documents', *Geographical Journal* 63 (1924), 391–412.
Hind, Arthur M. 'Queen Elizabeth by Augustine Ryther (?): An Undescribed State', *British Museum Quarterly* 9 (1934–5), 58.
Hind, Arthur M. *Engraving in England in the Sixteenth and Seventeenth Centuries. A Descriptive Catalogue with Introductions. Part 1. The Tudor Period*. Cambridge (1952), 85–95.
Hodson, Donald. *The Printed Maps of Hertfordshire 1577–1900*. Folkestone (1974).
Holinshed, Raphael and Harrison, William. *Chronicles* (1577–86).
Hull, F. *Catalogue of Estate Maps 1590–1840 in the Kent County Archives Office*. Maidstone (1973).
Hunter, Joseph. *South Yorkshire* (1828–31, reprinted Wakefield 1974).
Hunter, Joseph. Familae Minorum Gentium 2. Harleian Society 38 (1895).
Karslake, J. B. P. 'Further Notes on the Old English Mile', *Geographical Journal* 77 (1931), 358–60.
Kendall, H. P. 'The Story of a Local Feud', *Halifax Antiquarian Society's Papers* (1923).
Laxton, P. *Two Hundred and Fifty Years of Map-making in the County of Hampshire. A Collection of Reproductions of Printed Maps Published between the Years 1575 and 1826*. Lympne Castle (1976).
Leigh, Valentine. *The Most Profitable and Commendable Science of Surveying of Landes, Tenemens, and Hereditamentes* (1577).
Leybourn, William. *The Compleat Surveyor, Containing the Whole Art of Surveying of Land. By the Plaine Table, Theodolite, Circumferentor, Peractor, and other Instruments* (1653).
Lucar, Cyprian. *A Treatise named Lucar Solace* (1590).
Lynam, Edward. *Saxton's Atlas of England and Wales. An Introduction* (1936).
Lynam, Edward. 'The Development of Symbols, Lettering, Ornament and Colour on English Maps', *British Records Association Proceedings* 4 (1939), 20–34.
Lynam, Edward. *British Maps and Map-makers* (1944).
Lynam, Edward. 'English Maps and Map-makers of the Sixteenth Century', *Geographical Journal* 116 (1950), 7–28.
Lynam, Edward. *The Mapmaker's Art* (1953).
Manley, Gordon. 'Saxton's Survey of Northern England', *Geographical Journal* 83 (1934), 308–16.
Manley, Gordon. 'Observations on the Early Cartography of the English Hills', *Comptes Rendus du Congrès International de Géographie* 2. Amsterdam (1938), 36–42.
Marcombe, David. 'Saxton's Apprenticeship: John Rudd, a Yorkshire Cartographer', *Y.A.J.* 50 (1978), 171–5.
North, F. J. 'The Map of Wales', *Archaeologia Cambrensis* 90 (1935), 1–69.
Ogilby, John. *Britannia* (1675).
Parker, James. *Illustrated History from Hipperholme to Tong*. Bradford (1904).
Parsons, F. G. *The History of St Thomas' Hospital* (1932).
Peel, Frank. *Spen Valley Past and Present*. Heckmondwike (1893).
Phillips, P. L. *A List of Geographical Atlases in the Library of Congress*. Washington (1909–20).
Public Record Office. *Maps and Plans in the Public Record Office c.1430–c.1603*.
Rathborne, Aaron. *The Svrveyor in Foure Bookes* (1616).
Richeson, A. W. *English Land Measuring to 1800: Instruments and Practices*. Boston (1966).
Robertshaw, Wilfred. 'The Township of Manningham in the Seventeenth Century', *The Bradford Antiquary* 8 (1940).
Robertshaw, Wilfred. 'Notes on Adwalton Fair', *The Bradford Antiquary* 7 (1927).
Robertshaw, Wilfred. 'The Manor of Chellow', *The Bradford Antiquary* 9 (1940).
Robertshaw, Wilfred. 'A Saxton Family Note', *The Bradford Antiquary* 9 (1940).
Robertshaw, Wilfred. 'New Light on Adwalton Fair', *The Bradford Antiquary* 10 (1962).
Sargeant, W. J. 'A Further Discovery of Manuscript Maps by Christopher Saxton', *Geographical Journal* 132 (1966), 153–5.
Seebohm, Frederic. *Customary Acres and their Historical Importance* (1914).
Sheard, Michael. *Records of the Parish of Batley in the County of York*. Worksop (1894).

Skelton, Raleigh A. *Decorative Printed Maps of the 15th to 18th Centuries* (1966).
Skelton, Raleigh A. *County Atlases of the British Isles, 1579–1850. A Bibliography* (1970).
Skelton, Raleigh A. and Sommerson, John A. *A Description of the Maps and Architectural Drawings in the Collection Made by William Cecil, First Baron Burghley, now at Hatfield House*. Oxford (1971).
Skelton, Raleigh A. *Saxton's Survey of England and Wales: with a Facsimile of Saxton's Wall-map of 1583*. Amsterdam (1974).
Smith, A. G. R. *The Government of Elizabethan England* (1967).
Smith, R. B. *Land and Politics in the England of Henry VIII. The West Riding of Yorkshire: 1530–46*. Oxford (1970).
Somerville, Sir Robert. *History of the Duchy of Lancaster, Vol.1. 1265–1603* (1953).
Somerville, Sir Robert. *Office-Holders in the Duchy and County Palatine of Lancaster from 1603* (1972).
Stow, John. *The Survey of London* (1603).
Taylor, Eva G. R. 'The Earliest Account of Triangulation', *Scottish Geographical Magazine* 43. Edinburgh (1927), 341–5.
Taylor, Eva G. R. 'The Plane-table in the Sixteenth Century', *Scottish Geographical Magazine* 45. Edinburgh (1929), 205–11.
Taylor, Eva G. R. 'The Surveyor', *Economic History Review* 17 (1947), 121–33.
Taylor, Thomas. *The History of Wakefield: The Rectory Manor*. Wakefield (1886).
Tempest, Eleanor B. 'Broughton Hall and its Associations', *The Bradford Antiquary* 6 (1921).
Thoresby, Ralph. *Ducatus Leodiensis, or The Topography of the Town and Parish of Leeds and Parts Adjacent in the West-Riding of York* (1715).
Thoresby, Ralph. *The Diary of Ralph Thoresby 1677–1724*. Edited by Joseph Hunter. Leeds and Wakefield (1830).
Thrower, Norman J. W. *Maps and Man: An Examination of Cartography in Relation to Culture and Civilization*. Englewood Cliffs (1972).
Tooley, Ronald V. *Maps and Map-making* (1949).
Tyacke, Sarah. *London Mapsellers, 1660–1720* (1978).
Upton, Anthony F. *Sir Arthur Ingram c.1565–1642*. Oxford (1961).
Walker, E. J. Our Local Portfolio 143. *Halifax Guardian*, February 25th, 1860.
Walker, J. W. *Wakefield: Its History and People*. 2nd edition, Wakefield (1939).
Walpole, Horace. *Anecdotes of Printing in England* (1786).
Watson, John. *The History and Antiquities of the Parish of Halifax* (1775).
Wentworth, George E. 'History of the Wentworths of Woolley', *Y.A.J.* 12 (1893), 1–35, 159–94.
Whitaker, Harold. *A Descriptive List of the Printed Maps of Yorkshire and its Ridings, 1577–1900*. Leeds (1933).
Whitaker, Harold. *A Descriptive List of the Printed Maps of Lancashire, 1577–1900*. Manchester (1938).
Whitaker, Harold. 'The Later Editions of Saxton's County Maps', *Imago Mundi* 3. London (1939), 72–86.
Whitaker, Harold. *A Descriptive List of the Printed Maps of Cheshire, 1577–1900*. Manchester (1942).
Whitaker, Harold. *The Harold Whitaker Collection of County Atlases, Road Books and Maps Presented to the University of Leeds. A Catalogue*. Leeds (1947).
Whitaker, Harold. *A Descriptive List of the Printed Maps of Northamptonshire, 1576–1900*. Northampton (1948).
Whitaker, Harold. *A Descriptive List of the Maps of Northumberland, 1576–1900*. Newcastle-upon-Tyne (1949).
Whitaker, Thomas D. *Loidis and Elmete*. Leeds and Wakefield (1816).
Whone, Clifford. 'Christopher Danby of Masham and Farnley', *Thoresby Society* 37 (1936), 1–29.
Willan, T. S. 'Three Seventeenth Century Surveys', *Yorkshire Archaeological Society Record Series* 104 (1941), 1–79.
Worsop, Edward. *A Discoverie of Sundrie Errours and Faults Daily Committed by Landmeaters* (1582).

Index

Ackworth 79, 101, 122
Adwalton xvi, 2, 72–3, 141
Agas, Ralph 44
Agricola, Georg 42
Airmyn 122, 124, 126
Aldclose 79, 80, 81
Allison, Robert 81
Almshouses, at Westerton 71
Alnwick Castle 80, 116–18
Anglesey, map of 15, 17, 19, 28, 31, 38, 53, 60, 62, 64, 146, 155, 158
Anglia..., map of 9, 15, 20, 25, 27, 28, 29, 30, 31, 32, 34, 35, 36, 39, 40, 41, 45, 49, 52, 58, 59, 61, 62, 143, 149, 158
Arms, grant of 68, 69, 164
Armytage, Gregory 141, 142
Ashmole, Elias 68, 164
Ashwell 75, 78, 80, 115–16
Aveley 77, 79, 80–81, 84–5, 86
Axholme, Isle of 79, 99, 119

Bacon, Sir Francis 132–3
Baildon 75, 78, 117, 122, 125, 126
Baildon, Sir Francis 111, 126, 127
   William 127
Bailiff – Duchy of Lancaster 3, 70, 71, 166
   Land in London 67, 163
Ball, Edward 89
  Thomas 89
  William 88–9
Barber, Sir Robert 124
Barkisland 80, 97
Barton, Mr 124
Bate 85
Batt, Henry 71
Baylye, Thomas 124
Bayford 79, 85–6, 87
Bedfordshire, map of 10, 16, 18, 29, 30, 37, 49, 60, 61, 63, 144, 155
Beeston, Raphe 73
Belfast Lough 9, 120
Bellassis, Sir William 106
Bell, John 108
Belvoir 115
Benall 67, 147
Bennet, Constance 67, 163
Bennitland, William 132–3
Berkshire, map of 15, 18, 30, 37, 60, 62, 64, 144, 155
Bess of Hardwick 78, 114
Beverley 118
Bingley 88–9, 126
Binneman, Walter 61, 63

Birkbeck, Peter xvi, 4
Birkbye, Master 112
Birkenshaw 95
Birmingham Reference Library 35, 150
Birstall 2, 72, 136
Blackburn Public Library 162
Blades, Elizabeth 142
Blakey, John 135
Blois, Sir Gervase 91
  Martha 91
  Sir William 91
Blythburgh 91
Bodleian Library 6, 29, 45, 58, 68, 119, 132, 151, 161, 162, 163, 164
Bolling, Edward 88
  William 88
Boltby 106
Boothby, Sir Hugo 152
Borough, William 147
Borthwick Institute 1, 2, 89, 142
Botham Hall 80, 97
Bourne, William 43
Bradford – 72, 73, 88, 95, 114, 123, 126, 129, 130–31
  Library 135
Breconshire, map of 15, 17, 19, 30, 38, 62, 64, 146, 155
Brewer, Richard 132–3
Bridlington Priory 70, 166
Brindley, James 109
British Library 30, 35, 45, 52, 62–5, 143, 148, 149, 154, 162, 163
British Museum 79, 85, 120
Brograve, John 70, 167
Brooke, Alice xvi, 2, 72
  James 72
  John 2, 72–3, 142
  Martha 91
  Sir Robert 91
  William 72, 141
Brotherton Library 52, 79, 108, 109, 152, 163
Broughton 79, 80
Buckinghamshire, map of 15, 18, 30, 37, 60, 62, 64, 65, 144, 155
Burghley, Lord xiii, 7, 9, 10, 11, 20, 66, 70, 80, 120, 123, 127, 143, 144–6, 166
  proofs 9, 10, 12, 13, 14, 15, 16, 17, 20, 28, 29, 31, 32, 36, 39, 66, 143–6, 148
Burley in Wharfedale 80, 115, 117
Burnley 118–119
Burton, Bartholomew 116
  Lora 116
  Phillippa 124
  Sir Thomas 124
Byland 79, 103, 105–6

Caernarvonshire, map of   15, 17, 19, 38, 62, 64, 146, 155, 158
Caldecott, Henry   101
Calverley   123–4
Calverley, Henry   124, 130
    Mr   124
    Mrs   115
    Phillippa   124
    Sir Walter   124, 130
Cambridge – 80, 82, 83
    Fitzwilliam Museum   151
    Magdalene College   162
    Peterhouse   151
    University   5, 7
    University Library   45, 47, 64, 151, 162
Cambridgeshire, map of   16, 18, 30, 37, 60, 62, 144, 153, 155
Canterbury, Archbishop of   63, 52
Cardiganshire, map of   15, 17, 19, 30, 38, 60, 62, 64, 146, 155
Carey, Edward   71, 119
    Sir Phillip   120, 131
Carlton   129
Carmarthenshire, map of   15, 17, 19, 30, 38, 62, 64, 146, 155
Carpenter's square   42
Carter, John   86
Cartouche   30, 31, 47, 75, 85, 114
Cary, Robert   156
Casson, Roger   141, 143
Catlyn, William   83
Cattal   79, 110
Catton   118
Cavendish, Mary   78
    Sir William   78, 112, 114
Cecil, Robert   127
    William – see Burghley
Chadwick, William   143
Chatsworth   78, 79, 80
Chellow Common   88
Cherry Burton   118
Cheshire, map of   10, 14, 16, 18, 31, 37, 39, 45, 48, 49, 53, 58, 59, 60, 61, 63, 146, 155, 158
Chester Beatty Library   149
Chesterfield, Earl of   155
Chetham's Library   150
Chicago Newberry Library   153
Chronicles, by Raphael Holinshed   xiii, 11, 40, 66, 67, 68
Cities and Coat of Arms in Atlas   20, 26, 28, 149
Circumferentor   43, 44
Clapham, Charles   130
Clayton, Thomas   71
Clerkenwell   67, 69, 70, 165
Cleves, Lady Anne of   67, 147
Clifton, William   70, 166
Cockfield Hall   91
Colby, Mary   136
Cold Hiendley   80, 102, 108–9
Cole, George   136
    Thomas   136

Collins, Captain   64
Colne   134–5
Colouration of maps   34
Combe   79, 80, 81
Combe Grove   79, 80, 81
Compass   42, 43
Congress, Library of   153, 154, 163
Contents of Atlas   30–32
Copley, Alverey   5, 134, 135
    Edward   123, 134
    Elizabeth   134
    Isabel   5
    John   2
Cord   42
Cornwall, map of   9, 10, 11, 12, 13, 16, 18, 28, 30, 31, 37, 38, 39, 143, 155, 156, 158
Corker, James   101
Cosin, Christopher   4
Cosmographicall Glasse by William Cunningham   43
Cosmographie by Reynold Wolfe   11
Cowper, Daniel   123
    John   123
Crabtree, Nicholas   131
Cragfargus   120
Cranfield, Lionel   123
Crawford, Lord   152
Crofton   79, 80–82
Cross-Staff   42, 43
Crowder, Edward   71
Crowle   132
Crowther, Brian   124, 126
Cumberland, map of   13, 14, 16, 18, 50, 60, 61, 63, 145, 155
Cunliffe, Nicholas   135
Cunningham, William   43
Currer, W.   129

Danby, Christopher   127
Darrell, Cassandra   164
    Sir John   164
Dawney – archives   116
    John   116
    Lora   116
Daywork – see Measurements
Deane, Robart   94
Dee, John   5, 43, 100
de Hooghe, Cornelis   15, 35, 39
Denbighshire, map of   14, 15, 16, 19, 28, 31, 38, 39, 45, 53, 60, 62, 64, 146, 155, 158
Dent, Joseph   111
Description of Britaine by William Harrison   11, 40, 67
Derbyshire, map of   14, 16, 18, 31, 37, 38, 45, 48, 49, 59, 60, 61, 63, 145, 155, 159
Devonshire, Duke of   112
    map of   9, 12, 16, 18, 30, 31, 37, 38, 39, 48, 53, 59, 60, 62, 63, 143
Dewsbury   xvi, 1, 4, 5, 6, 79, 96, 111, 131
    – Library   111
Dicey, C. & Co; edition of *c.*1770   58
Digges, Leonard   42, 43
Doncker, Hendrik   52

Dorset, map of   9, 12, 16, 18, 30, 37, 39, 45, 48, 49, 59, 60, 62, 63, 143, 155, 159
Downe, Viscount   116
Drighlington   72, 95
Dublin University, Trinity College Library   152
Duchy of Lancaster, bailiff   3, 70–71, 166–7
      land   4, 5, 72, 99, 101, 105, 107, 118, 119, 120, 122, 123, 129, 134–5, 166–7
Dudlington   101
Dunningley   1, 2, 4, 69, 71, 72, 99, 100, 119, 136, 141, 142, 164, 147
Durham, Cathedral   6
   map of   13, 14, 16, 18, 31, 37, 39, 45, 48, 49, 60, 61, 63, 145, 155, 156, 159
Durnford, E. W.   93, 94

East Bierley   79, 93, 95–6
Egerton, Sir Thomas   102
Elland   4, 77, 80, 93, 97
Elland, Isabella   95
  Sir Thomas   95
Elland Park   79, 92, 93, 94–5
Ellington Moor   122, 127
Elwys, John   114
Emley   80, 97
Enclosure   96, 102, 108, 116, 118, 131, 133, 135
England and Wales, map of   29, 34, 35, 36, 41, 61 (see also Anglia)
Engravers   12, 14, 31, 35–6, 39
'Epitaph'   2, 6–7, 66
Esholt   122, 126, 128, 129, 130
Essex   74, 79, 80, 81, 83, 84
 – Arthur Earl of   61, 63, 64
  map of   10, 11, 13, 15, 16, 18, 31, 37, 39, 45, 48, 52, 59, 60, 61, 63, 144, 155, 159
Exchequer Court   67, 70, 80, 99, 105, 106, 136, 166
Exeter, Lord   116
Exley Hall   94
Extwistle   118

Fairbank, George   71, 72
Falmouth   9, 143
Fanscombe   79, 80, 81
Farnley Hall   127
Farrar, Henry   90, 107–8
Farsley   122, 123–4
Faversham   79, 85–7
Favour, Dr John   2, 6, 7, 66, 126
Fernaham   67
Ferres, William   155
Field, John   5, 88
Finch, Thomas   86
Firth, Richard   67
Fitzwilliam, Gervase   126–7
Fitzwilliams, Michal   120
Flintshire, map of   14, 15, 16, 19, 38, 62, 64, 146, 155
Flower, William   68, 164, 165
Folger Shakespeare Library   79, 153
Foljambe family   102

Foster, William   85
Foulis, Henry   6
Foullon, Abel   42
Foxcroft, John   70–71, 166, 167
  Michael   107–8
Freeman, J.   84, 85
French edition of atlas   50, 52
Frobisher, Martin   101
Frontispiece   20, 21, 25, 29, 30, 36, 149
Fulbourn   79, 80, 82–3

Gardner, John   163
Gargrave, Sir Richard   123
  Sir Thomas   123
Garrard, Sir John   86
  Sir William   86, 87, 167
Garrett, John   155
Gascoigne, Henry   143
Gawdy, Henry   91
Geometrical square   42
Gerons   79, 80, 83
Gerrard (see also Garrard)   70, 71, 105, 167
Glamorgan, map of   15, 17, 19, 28, 31, 38, 39, 41, 45, 48, 60, 62, 65, 146, 155, 156, 159
Glasgow University Library   152
Glenham, Great and Little   67, 147
Glenham, Thomas   67
Gloucestershire, map of   14, 16, 18, 23, 31, 32, 37, 38, 39, 45, 48, 49, 53, 59, 60, 62, 63, 146, 155, 156, 159
Glover, Agnes   141
  Anthony   141, 142
  Robert   68
Golthwaite, Elizabeth   143
  Robert   143
Gombsall, Elizabeth   136
  John   136
Gomersal   74, 75, 96, 122, 136
Goodman, John   136
Goodmansterne   79, 85, 87
Goodricke, Sir Henry   111
  Sir John   109, 111
  Margaret   111
  Richard   78, 110, 111
Gough Map   xiii
Gray's Inn Library   154
Great Parndon   83
Greenwood's Hospital   71
Greenwood, James   71
  Jane   141
  Robert   71, 141
Grey, Dame Isabel   164
Grice, Henry   90
Grigston (Griston)   11, 12, 67, 147
Grimwood, Thomas   91
Guernsey   37, 62, 64

Haddlesey   79, 112, 114
Haigh, John   134, 142
Halifax   6, 7, 93, 94, 100, 107, 119, 124, 126, 131
Hall, Thomas   71, 167
  William   72–3

181

Hampshire, map of   12, 16, 18, 29, 30, 31, 32, 33, 37, 38, 39, 41, 48, 49, 53, 59, 60, 62, 63, 64, 144, 155, 156, 159
Hanson, John   119, 131, 132
Harewood, Earl of   111
Harrison, William   11, 40, 60, 130
Hartley, Prudence   xvi, 3, 141
Harvard University, Houghton Library   154
Hasell   118
Hastingleigh   79, 80, 81
Hatfield   122, 132–3
Hatfield Chace   99, 133
Hatton, Sir Christopher   28
Hawtayne, Edward   67
Haxey   99, 100
Heath Grammar School   124, 126
Heaton Common   79, 88, 130
Heaton, Robert   135
Hely, Richard   99
Heneage, Sir Thomas   28, 144
Herdson, John   84
Herefordshire, map of   14, 16, 18, 37, 39, 45, 47, 48, 49, 59, 60, 61, 63, 146, 155, 156, 159
Hertfordshire, map of   10, 11, 14, 17, 18, 28, 30, 31, 36, 37, 39, 59, 60, 61, 64, 144, 155, 159
Hey, Robert   135
Hicks, Sir Baptist   70
Hinton   79, 91, 93
Hinton Netherhall   79, 80, 83
Hirst, Thomas   72, 141
Hobson, Alice   xvi, 2, 142
       Mary (Marie)   xvi, 2, 3, 142, 143
Hodgson, Richard   88
       William   88
Hodgson, D.   163
Hogenberg, Frans   35
       Remigius   12, 16, 17, 20, 35, 36, 39
Holinshed, Raphael   xiii, 11, 40, 66, 68
Holometer   42
Homestall Farm   87
Hondius, Henricus   60, 153
       Jodocus   154, 155
Honour of Pontefract   5, 71, 101, 105, 129
Hopton, Sir Arthur   91
       Sir Owen   91
Horbury   77, 79, 93, 96, 102, 105, 141
Hose   101
Houghton Library   154
Howley Hall   4, 5, 71, 124, 127, 134
Hundreds   10, 11, 12, 13, 15, 16, 17, 33, 34, 49, 53, 155, 156, 158, 161
Hunsingore   75, 79, 109, 110, 111
Hunsworth   79, 93, 95–6
Hunt, Nicholas   133
Huntingdonshire, map of   16, 18, 30, 37, 49, 60, 61, 63, 144, 155
Huntingdon, Henry E., Library   153

Illinois University Library   154
Index, in atlas   20, 22, 23, 24, 25, 28, 30, 36, 149
Ingram, Sir Arthur   123, 124, 126

Inner Temple Library   150
Intermediate states   57, 58
Ireland   30, 41, 60, 65, 120, 154
Isle of Axholme, see Axholme
Isle of Man, map of   41, 64
Isle of Wight   9, 41, 60, 62, 64, 144, 155
Itinerary by John Leland   40, 41

Jackson, Charles   85
       John   120
Jeffreys, Thomas; edition of c.1749   58
Jersey   10, 37, 62, 64, 146
Johnson, Henry   111
       Family   111
       Frances   111
Judson, John   118

Kaye, Jonathan   132
Keighley   80, 112, 114, 134
Keighley, Ann   114
       Henry   114
Kelfeilde   100
Kent   74, 75, 79, 80, 81, 85, 87
Kent, map of   12, 13, 16, 18, 25, 28, 30, 31, 32, 36, 38, 39, 45, 52, 53, 59, 60, 62, 63, 64, 144, 155, 160
Kent County Archives   79
Kildwick   134
Kirby   106

Lacey, John   94
Lamb, Francis   62, 63, 159
Lambarde, William   40
Lambeth Palace Library   154
Lampe, William   118
Lancashire   72, 80, 100, 118–19, 134
Lancashire, map of   10, 11, 14, 17, 18, 31, 32, 33, 37, 38, 39, 45, 47, 48, 49, 60, 61, 63, 146, 155, 156, 160
Lascelles family   111
Lea, Philip; edition of c.1689   47–8, 49, 50, 51, 52
       maps   61–2, 158–62
       edition of c.1693   47–8, 49, 50, 51, 52, 53, 58
       maps   62–5, 158–62
Leconfield   118
Ledston   109, 122, 133–4
Lee, Thomas   93, 94
Leeds   1, 4, 5, 36
   – Central Library   150
   City Archives   79, 89, 111, 122, 123, 126, 129, 133, 134
   University (see Brotherton Library)
Leicestershire   101
Leicestershire, map of   16, 18, 37, 60, 61, 64, 145, 155
Leigh, Valentine   43
Leighton Hall   127
Leland, John   40, 41
Leslie, John   146
Leybourn, William   43, 44
Lhuyd, Humphrey   xiii, 41, 151

Library of Congress   153, 154, 163
Licence to print   11, 14, 15, 68, 147
Lincolnshire   77, 99, 132
Lincolnshire, map of   13, 16, 18, 31, 32, 33, 34, 37, 38, 39, 45, 47, 49, 59, 60, 61, 63, 145, 155, 160
Linley(e), Alveraye   71
    Dorothy   142
    Family   4, 72
    Jasper   141, 142
    Nicholas   71, 142
    Robert   142
Linnyall   101
Linton   80, 116, 118
Liverpool University, Harold Cohen Library   152
Liversedge   80, 119–20, 122, 131
London (Greater) Record Office   79
    Library   150
    Society of Antiquaries of   150
Lucar, Cyprian   43
Luddenden   79, 106–108
Lupset   95, 114, 115

Manchester   5, 79, 100
Manchester Public Library   47, 49, 63, 64, 163
    University, John Ryland's Library   152
Manningham   74, 75, 78, 122, 130–31
Manwood, Sir Roger   80
Marberie   101
Masham   127
Mawe, Robert   71, 167
Measurements, land   42–4, 69, 74, 75, 90, 96, 110, 136, 167
Medley   88
Medlow   79, 87
Mercator, Gerard   xiii, 6, 34, 35, 40, 41, 100
Meringe, Marion   164
Merionethshire, map of   15, 17, 19, 38, 62, 64, 146
Metcalfe, Christian   164
Methley   75, 79, 87, 88, 89–91
Middlesex   67, 70, 79, 80, 163, 165
Middlesex, map of   16, 18, 30, 36, 60, 61, 63, 144, 155
Middlestown   80, 97
Middleton, Sir Peter   115, 117
Midgley   93, 94, 107
Midgley, John   124, 126, 135
Mildmaye, Sir Walter   70, 80, 166
Milton   144
Misterton   90, 100
Mitchell, Christopher   135
Moll, H.   65
Monastic lands   70, 74, 80, 85, 89, 130, 166
Monk Bretton   79, 85
Monmouthshire map of   14, 17, 18, 28, 38, 39, 45, 48, 49, 60, 62–3, 146, 155, 156, 160
Monnsaughes Farm   83
Monnsaughes, William   83
Monteagle, Lord   127
Montgomeryshire, map of   15, 17, 19, 28, 32, 38, 39, 53, 60, 62, 64, 146, 155, 160
Moore, Sir Jonas   62

Morgan, William   61, 63
Morley wapentake   4, 47, 70, 166
Morton   106

Nalson, Alice   141
    Grace   xvi, 3, 4, 143
    Mary   4, 143
    Nicholas   141
    Thomas   xvi, 3, 4, 141, 143
National Maritime Museum   150, 163
Netherton   79, 102, 105
Nettleton, John   142
Newbie   100
Newcastle-upon-Tyne City Library   163
Newehall   96
Newport Public Library   163
New York   101, 122, 153
Nicholls, Sutton   58
Nicholson, E. R. A.   154
Norden, John   153
Norfolk, map of   10, 11, 12, 15, 18, 28, 29, 30, 31, 32, 33, 35, 37, 38, 39, 47, 48, 49, 58, 60, 61, 63, 144, 153, 155, 156, 160
Northamptonshire   116
Northamptonshire, map of   10, 13, 16, 18, 28, 29, 30, 31, 36, 37, 38, 45, 47, 48, 49, 54, 55, 58, 59, 60, 61, 63, 144, 149, 155, 157, 160
North Bierley   95
Northropp, William   131
Northumberland, Earl of   78
Northumberland, map of   10, 13, 14, 17, 18, 25, 30, 32, 36, 38, 45, 48, 49, 53, 56, 57, 58, 60, 61, 64, 145, 160
North Yorkshire Record Office   80, 116
Norway   147
Norwich, Dean and Chapter of   70
Nottinghamshire, map of   13, 16, 18, 37, 59, 60, 61, 63, 145, 155
    Record Office   79, 93, 94, 95, 96, 97, 102, 122
Notton   79, 108–9
Nowell, Laurence   40
Nutter, Anthony   143

Oakham   116
Oakworth   122, 134–5
Ogilby, John   50, 53, 61, 63, 162
Oldfield   123–4
Oliver, John   64
Ortelius, Abraham   xiii, 35, 40, 146
Osboltson, Lambert   81–3
Osberton Hall   79, 102
Ossett   xvi, 1
Otley   115
Owston   99–100
Oxenford, Earl of   84
Oxenhope   93
Oxfordshire, map of   10, 11, 12, 15, 18, 28, 30, 31, 32, 36, 37, 38, 45, 53, 59, 60, 62, 64, 144, 155, 161

Oxford University
　　– Balliol College Library　151
　　Bodleian Library (see Bodleian
　　　Library
　　Christ Church Library　151, 154
　　New College Library　151
　　Queen's College Library　152
　　Worcester College Library　163

Page, Margaret　xvi, 2, 3, 142
　　Matthew　xvi, 2, 3, 142
　　Peter　xvi, 2, 3, 72, 142, 167
　　Thomas　2
Palmer, Richard　64
Palmes, Francis　116
Pantometria by Leonard Digges　43
Paris, Matthew　xiii
Parishes　10, 11
Parker, Frances　127
　　John　118–19
Payment　82, 118
Peck, Joseph　71
Pepper, Joseph　116
Percehay, Anthony　165
　　Richard　165
　　Walter　165
　　William　165
Percy, Henry　117, 118
　　Thomas　117
Petty, Sir W.　65
Phillips, Sir Thomas　101, 112, 122
Phrysius, Gemma　43, 100
Pierpont Morgan Library　79, 122, 153
Pierse, Samuel　136
Pighell, Henry　135
Pembrokeshire, map of　15, 17, 19, 28, 38, 39, 53, 60, 62, 64, 146, 155, 161
Pilkington, John de　115
　　Sir John　115
　　Thomas　114–15
Plane-table　42, 43, 44
Pole　42
Pontefract　4, 5, 80, 101, 127, 129, 133
　　– Honour of　5, 71, 101
　　West　70, 166
Popeley, Francis　137
　　Grace　137
　　John　137
Powell, Mr　93, 94
Priestethorpe　123–4
Priestthorpe　75, 79, 88–9, 126
Public Record Office　2, 79, 80, 85, 87, 88, 89, 105, 106, 107, 112, 118, 120, 121, 122, 127, 130, 132, 134, 147
Pulleyn, Francis　115
Purfleet　84
Purston, Edward　71

Rabbes Farm　79, 80, 83
Radnorshire, map of　15, 17, 19, 28, 30, 36, 38, 53, 60, 62, 64, 146, 155, 161

Ramsden, Anthony　4
　　Elizabeth　xvi, 3, 4, 143
　　Isabel　4
　　John　xvi, 4
Rathborne, Aaron　43, 44
Rawson, Arthur　135
Rawtenstall　80, 97
Rayner, Richard　131
Rea, Roger　154, 155
Regiomontanus　42
Reisch, Gregor　43
Rendham　67, 147
Reyner, John　143
Reynolds, Nicholas　17, 36, 39
Ribston　75, 79, 109–11
Richardson, Bryan　124
Rievaulx Abbey　106
Robert-Town　120
Rouland, Arthur　100
Royal Arms (of Elizabeth I)　12, 15, 28, 31, 34, 45, 47, 53, 58, 66, 158, 159, 160, 161
Royal Geographical Society　28, 47, 50, 61, 62, 150, 162
Roze (Rotz), Jean　43
Rudd, John　1, 6, 7, 139
Rufford Abbey　94, 95
Rushworth　80, 97
Rutland　80, 115–16
　　Earl of　116
　　map of　16, 18, 30, 37, 49, 60, 61, 63, 144, 155
Rutlinger, Johannes　15, 17, 36
Ryther, Augustine　14, 15, 16, 17, 20, 28, 30, 32, 35, 36, 39, 149, 158

St John of Jerusalem　67, 163, 165
St. Sepulchre without Newgate　69, 165
St. Thomas' Hospital　74, 77, 78, 81–5, 86
Saker, Henry　87
Salop Record Office　79, 101
Sandal　3, 75, 101, 122
Sandal Castle　72, 119
Savage, Mr　129
Savile, Edward　95, 96, 114
　　Elizabeth　5, 95, 134
　　Family　5, 78, 94, 102, 114, 123
　　Sir George　78, 93, 94–5, 99, 114, 129
　　Henry　5, 95, 96, 100
　　Isabella　95
　　John　4, 5, 71, 90, 95, 112, 124, 127, 134, 151
　　Lady　114
　　Sir Robert　5
　　William　114
Savile Estate Office　80, 97
Savill, Mr　123
Savoy Hospital　81, 82–3
Saxton, Agnes　164
　　Alice　xvi, 1, 2, 142
　　Ann　xvi, 3, 73, 141, 142
　　Anthony　139
　　Brian　164, 165

    Christopher (son of surveyor) 2, 3, 142, 143
    Elizabeth xvi, 3, 4, 143
    Francis xvi
    Grace xvi, 3, 4, 143
    Hugh 165
    Jenet xvi
    John xvi, 1, 141, 164
    Margaret xvi, 2, 71
    Margarete xvi, 1
    Margery xvi, 1, 141
    Mary xvi, 3, 4, 141, 143, 165
    Mrs 3, 141, 142
    Nicholas 164, 165
    Prudence xvi, 3, 141
    Rauf xvi, 1
    Robert xvi, 1, 2, 3, 4, 5, 72–3, 74–5, 78, 99, 101, 109, 114, 117, 120, 122–39, 141, 142, 164, 165
    Thomas xvi, 1, 2, 3, 72, 134, 141, 142, 143
    William xvi, 3, 141, 143, 164, 165
Scarborough 70, 145
Scarbrough, Lord 93, 94
Scalby (see Scawby)
Scales, atlas 32, 38–9
    manuscript maps 75, 76
Scatter, Francis 16, 17, 36, 39, 158
Scawby 70, 166
Scorbrough 118
Scotland 10, 11, 30, 41, 60, 65, 146, 147, 154
Seckford – Thomas 7–8, 11, 40, 66, 67, 68, 147, 163
    motto 11, 12, 13, 15, 17, 31, 34, 39
    Saxton's patron xiii, 7, 31
Selby 114
Seller, John 50, 63, 64, 65
Senior, William 112
Shafton 76, 77, 79, 101, 102
Shrewsbury, Earl of (see Talbot)
Shropshire – 79, 101
    map of 10, 14, 17, 18, 31, 37, 39, 45, 47, 48, 49, 53, 59, 60, 61, 64, 145, 146, 155, 157, 161
Sicklinghall 115, 117
Signet Library 154
Singleton, John 70, 166
Sion College 154
Sittingbourne 86
Skelmanthorpe 80, 97
Skircoat 124
Skipton 80
Snaith 124
Snapethorpe 3, 75, 80, 113–15
Somerset, map of 12, 13, 16, 18, 28, 30, 31, 37, 38, 39, 45, 47, 48, 49, 50, 53, 59, 60, 62, 64, 144, 155, 157, 161
Somerset, Thomas 120–21
Southowram 96
Sowood 1
Special Commission 80, 99, 133, 135, 136
Speed, John 139, 154, 155, 156, 157, 160, 161

Spofforth 80, 116–18
Stafford 67, 147
Staffordshire, map of 14, 17, 18, 31, 37, 38, 39, 47, 48, 49, 59, 60, 61, 64, 145, 155, 157, 161
Stainland 4, 80, 97
Stamford 116
Stanhope, Edward 108
Stanbury 135
Standley, Thomas 155
Stanley Hall 115
Stansfield, Robert 130
Steanor 96, 105, 112
Stevens, Dorothy xvi
Stevenson, Robert 133
Steward 79, 80, 83
Stockeld Park 115, 117
Stockwith 99
Suffolk – 2, 7, 31, 71
    grant of land 11, 67, 147
    map of 10, 11, 12, 13, 15, 16, 18, 28, 30, 31, 36, 38, 39, 45, 47, 48, 49, 59, 60, 61, 64, 144, 153, 155, 157, 161
    Record Office 79, 91
Sutton in Craven 122, 134
Surrey, map of 16, 30, 36, 60, 62, 64, 144, 155
Surveying instruments 42
Sussex, map of 16, 18, 30, 32, 36, 41, 60, 62, 64, 144, 155
Swefflinge 67, 147
Swinden, river 80, 118–19

Tadcaster 139
Tailor, Robert 85
Talbot, Edward 78, 101, 102
    George 78, 95, 101, 102
    Gilbert 71
    Grace 78
    Henry 101
    Mary 78, 95, 102
Talerferes 79, 80, 83
Tanshelf 74, 101, 122, 127, 129
Tectonicon by Leonard Digges 42, 43
Tempest, Henry 80, 96
    Steven 80
Taylor, William 124
Tartaglia, Niccolo 42, 43
Terwoort, Lenaert 12, 13, 16, 17, 28, 31, 35, 36, 39, 162
Teversham 79, 80, 82–3
Thewe, John 118
Theodolite 43, 44
Theodolitus (of Leonard Digges) 43
Thomas, Mary 88
    William 88
Thompson, Henry 130
    Frances 130
Thompsonn, Mary xvi, 4, 141
Thoresby, Ralph 36, 111
Thorne Waste 132
Thornhill 5, 6, 74, 75, 77, 80, 93, 95, 96–9, 111, 112, 122, 129

Thornhill, Elizabeth 95
  Simon 95
Thwaites, Katherine 164
Tingley (Tynglawe) 1, 71, 141
Tirell, Charles 131
Tockledge 132
Tomsons Farm 120
Tong 73, 95, 96
Topcliffe 71, 141
Topographical Dictionarie by William Lambarde 40
Topographical Instrument (of William Bourne) 43
Towneley, John 118—19
Triangulation 43-4

Ubaldini, P. 28, 30, 149
Universal Cosmographie by Reynold Wolfe 40-41

Vavasour 117
  Sir Mauger 126, 127
Vermuyden 133

Waddington, Christian 142
  Leonard 142
Wade, Robert 88
  Samuel 107
  Thomas 91
Wadlands 123-4
Wadsworth 78, 79, 80, 93, 94, 95, 97
Wakefield 1, 3, 4, 5, 6, 7, 95, 105, 114, 115, 122
  – Manor of 5, 7, 71-2, 102, 112, 119, 122
  Museum 79, 108
  Old Park 122, 123
  Outwood 120-21
  Queen Elizabeth Grammar School 5
Waldseemuller, Martin 43
Wales, National Library of 149, 163
  University College of North Wales Library 152, 162
  order of assistance 11, 14, 66, 147
Walker, Anthony 89
Walshford 79, 109
Walsingham, Sir Francis 70, 71, 153, 167
Walton Head 79, 109, 111
Wapontake, Agbrigg 4, 47, 70, 166
  Morley 4, 47, 70, 166
Warley 93, 107
Warner, Henry 91
Warwickshire, map of 13, 16, 18, 30, 31, 37, 38, 39, 53, 59, 60, 61, 64, 145, 155, 161
Washington DC, Library of Congress 153
Waterhouse, Edward 124, 126
Watermarks 29, 35
Watersheddles Reservoir 135

Waywiser 43
Web, William – edition of 1645 34, 45, 46, 47, 48, 49, 58, 155, 158, 159, 160, 161, 162
  titles of maps 59-60
Wensleydale 3, 122, 136
Wentworth, Michael 78, 102, 108-9, 127
  Sir Thomas 133-4, 137
West Ardsley 1, 2, 3, 99, 141, 142
Westerton 71
Weston tapestry maps 41
Westmorland, map of 13, 14, 16, 18, 31, 32, 33, 37, 38, 39, 45, 49, 50, 53, 60, 61, 63, 145, 155, 162
Wheatley, Thomas 141
Whitaker, Harold; projected edition of 1665 47, 48, 49, 155, 159
Whit(e), Thomas 95
Whitchurch 74, 79, 101
Wholeye, Thomas 71
Wilbfosse 118
Wilbore, Edmund 166, 167
Wilkinson, Richard 131
Willdey, George; edition of *c.*1730 53, 58, 163
Williams, John Camp 154
Wilson, Marmaduke 80
Wiltshire, map of 13, 16, 18, 30, 31, 37, 39, 45, 47 48, 49, 50, 53, 59, 60, 62, 64, 144, 155, 162
Wise, Mr 82
Witham, Henry 133-4
  Mrs 133-4
  William 133
Wolfe, Reynold xiii, 11, 40, 41
Wolstenholm, Sir Thomas 61, 63
Woodkirk (Woodchurch) 1-5, 71, 141, 142, 143
Woolley 79, 108, 127
Worcestershire, map of 14, 17, 18, 31, 37, 39, 45, 48, 49, 53, 59, 60, 61, 64, 145, 155, 157, 162
Worrell, Hugh 124
Worsop, Edward 42
Worsthorne 118
Wortley, Richard 85
Wotton, Sir Edward 106
Wycoller 134-5

York, Edwin, Archbishop of 90
  John, Archbishop of 90
  Minster Library 150, 164
Yorkshire Archaeological Society 80, 109, 111, 115, 122, 123, 126, 129, 136
Yorkshire, map of 11, 14, 17, 18, 28, 30, 31, 32, 34, 37, 38, 39, 40, 45, 47, 48, 49, 52, 60, 61, 64, 145, 155, 157, 162
Young, John 90
Younge, Charles 164
  Richard 164